Speaking of Genetics

A COLLECTION OF INTERVIEWS

ALSO FROM COLD SPRING HARBOR LABORATORY PRESS

A Passion for DNA: Genes, Genomes, and Society by James Watson

Davenport's Dream: 21st Century Reflections on Heredity and Eugenics edited by Jan A. Witkowski and John R. Inglis

Dorothy Hodgkin: A Life by Georgina Ferry

Francis Crick: Hunter of Life's Secrets by Robert Olby

George Beadle, An Uncommon Farmer: The Emergence of Genetics in the 20th Century by Paul Berg and Maxine Singer

I Wish I'd Made You Angry Earlier: Essays on Science, Scientists, and Humanity by Max F. Perutz

Max Perutz and the Secret of Life by Georgina Ferry

Mendel's Legacy: The Origin of Classical Genetics by Elof Axel Carlson

Sydney Brenner: A Biography by Errol C. Friedberg

The Eighth Day of Creation: The Makers of the Revolution in Biology (25th Anniversary Edition) by Horace Freeland Judson

The Writing Life of James D. Watson by Errol C. Friedberg

We Can Sleep Later: Alfred D. Hershey and the Origins of Molecular Biology edited by Franklin W. Stahl

What a Time I Am Having: Selected Letters of Max Perutz edited by Vivien Perutz

Speaking of Genetics

A COLLECTION OF INTERVIEWS

Jane Gitschier

University of California, San Francisco

Cold Spring Harbor Laboratory Press
Cold Spring Harbor, New York • http://www.cshlpress.com

Speaking of Genetics: A Collection of Interviews

© 2010 by Cold Spring Harbor Laboratory Press, Cold Spring Harbor, New York
All rights reserved
Printed in the United States of America

Publisher	John Inglis
Acquisitions Editor	Alexander Gann
Development Director	Jan Argentine
Project Manager	Joan Ebert
Production Manager	Denise Weiss
Production Editor	Mala Mazzullo
Typesetter	Techset Composition Limited
Desktop Editor	Susan Schaefer
Interior Book Designer	Denise Weiss
Cover Designer	Ed Atkeson

Front Cover: A page from the notebook of Sir John Sulston.

Library of Congress Cataloging-in-Publication Data

Gitschier, Jane.
 Speaking of genetics : a collection of interviews / Jane Gitschier.
     p. ; cm.
 Includes bibliographical references and index.
 ISBN 978-1-936113-03-3 (pbk. : alk. paper)
  1. Geneticists--Interviews. 2. Genetics--History--Sources. I. Cold
Spring Harbor Laboratory. Press. II. PLoS genetics. III. Title.
 [DNLM: 1. Genetics--Interview. QU 21]

 QH26.G58 2011
 576.5--dc22

2010026037

10 9 8 7 6 5 4 3 2

Authorization to photocopy items for internal or personal use, or the internal or personal use of specific clients, is granted by Cold Spring Harbor Laboratory Press, provided that the appropriate fee is paid directly to the Copyright Clearance Center (CCC). Write or call CCC at 222 Rosewood Drive, Danvers, MA 01923 (978-750-8400) for information about fees and regulations. Prior to photocopying items for educational classroom use, contact CCC at the above address. Additional information on CCC can be obtained at CCC Online at http://www.copyright.com/.

All Cold Spring Harbor Laboratory Press publications may be ordered directly from Cold Spring Harbor Laboratory Press, 500 Sunnyside Boulevard, Woodbury, New York 11797-2924. Phone: 1-800-843-4388 in Continental U.S. and Canada. All other locations: (516) 422-4100. FAX: (516) 422-4097. E-mail: cshpress@cshl.edu. For a complete catalog of Cold Spring Harbor Laboratory Press publications, visit our World Wide Web Site http://www.cshlpress.com/.

The Interviews presented in this volume were first published on the World Wide Web by the Public Library of Science (PLoS) under a Creative Commons Attribution License. Please visit www.plosgenetics.org for more information.

To Annie

I think, at a child's birth, if a mother could ask a fairy godmother to endow it with the most useful gift, that gift should be curiosity.

—Eleanor Roosevelt

Contents

Preface, ix

Victor Ambros	In the Tradition of Science, 1
Adrian Bird	On the Track of DNA Methylation, 11
David Botstein	Willing to Do the Math, 21
Herb Boyer	Wonderful Life, 31
Pat Brown	You Say You Want a Revolution, 43
Rebecca Cann	All About Mitochondrial Eve, 53
Sean Carroll	Curling Up with a Story, 63
Tom Cech	Meeting a Fork in the Road, 75
Soraya de Chadarevian	Twenty Paces from History, 83
Evan Eichler	Stable in a Genome of Instability, 93
Jenny Graves	The Exception That Proves the Rule, 105
Sir Alec Jeffreys	The Eureka Moment, 115
Judge John E. Jones III	Taken to School, 127
Mary Lyon	The Gift of Observation, 141
Svante Pääbo	Imagine, 153
Neil Risch	The Whole Side of It, 163
Elaine Strass	Ready for Her Close-Up, 171
Sir John Sulston	Knight in Common Armor, 181
Jamie Thomson	Sweating the Details, 199
Shirley Tilghman	The Making of a President, 211
Nicholas Wade	Turning the Tables, 221
Spencer Wells	Off the Beaten Path, 231

Further Reading, 243

Index, 247

About the Author, 259

Preface

Several years ago I was invited to write a historical piece about a project I had worked on two decades earlier. It was a time-consuming but meaningful adventure of conjuring up lost memories, digging through old lab notebooks, re-examining antiquated papers, dusting off slides and faded photographs, and ultimately committing to words the exhilaration of doing science. When I completed the task, an idea started to percolate: how could I tweak other scientists into telling their own stories of discovery? The answer soon came when the journal *PLoS Genetics* was founded and Mark Patterson, the director of publications for the Public Library of Science, asked me if I would consider writing interviews for it. With no experience and almost no hesitation, I said, "Yes!"

These interviews are my attempt to capture the essence of doing science as practiced or witnessed in the field of genetics. My intention has been not only to document important moments of scientific discovery, but also to convey the excitement, the labor, the serendipity, and the human joy and anguish inherent in it. I hope you will agree that the stories are inspirational, often humorous, and always compelling.

"How do you choose whom to interview?" I am often asked. I have attempted to meet with people who have worked on a variety of organisms and a range of scientific problems. I looked for people who achieved success accidentally as well as those who made a focused assault toward a discovery. I also chose subjects whose breadth of experience transcended the laboratory—the president of a university, an author, the founder of a company, and an explorer—to gain a wider perspective. Moreover, I was interested in expanding the dialog to include a set of individuals whose work interfaces with genetics, but who are not themselves scientists: a judge, a reporter, an administrator, and an historian of science. It was remarkable to me that every person I approached, with only one exception, readily agreed to be interviewed.

The interview "protocol" I established was simple, and I communicated it to the interviewees at the outset. With two exceptions, I met my subject in person, usually on his or her home turf. I recorded the entire interview, except when asked to turn off the recorder, and I transcribed the bulk of it. Then began the fun, yet challenging, process of redacting the interview into a cohesive product of only about 4000 words, including a title and a brief introduction. I sent this draft to the interviewee to check for accuracy and approval. The editors at *PLoS Genetics* also provided extremely valuable input, especially in helping me to cut parts that I couldn't bear to discard on my own. I also added a photo, usually one that I had taken on the spot. [Others were supplied by Victor Ambros (self-portrait), Rebecca Cann (courtesy of Lenny Freed), Tom Cech and Evan Eichler (courtesy of Howard Hughes Medical Institute), Svante Pääbo (courtesy of Max Planck Institute for Evolutionary Anthropology), and Nicholas Wade (courtesy of *The New York Times*).]

I thank my friend Mark Patterson for offering me the opportunity to write these interviews, all of which were conducted and originally published in *PLoS Genetics* during the period 2005–2010. The journal's first editor-in-chief Wayne Frankel was an enthusiastic and alert sounding board, and managing editor Catherine Nancarrow has been more cheerful and consistently supportive than anyone could hope for; ditto for Andy Collings in the Cambridge UK office. The other *PLoS* academic editors, staff editors and copy editors who have helped me over these five years are now quite numerous, and thanks go to Nicole Sheikh, Johanna Dehlinger, Tamsin Milewicz, Sam Moore, Catriona Silvey, Lizzy Fisher, Susan Rosenberg, and the two Gregs (Copenhaver and Barsh). The summer course at the UC Berkeley Regional Oral History Office provided me with the confidence and some special tips to move forward, and I especially thank Sally Smith Hughes for her encouragement and insight. I have been very lucky to work with Alex Gann, Joan Ebert, Jan Argentine, Mala Mazzullo, and Denise Weiss at Cold Spring Harbor Laboratory Press, and I thank Sandy Johnson for making that connection. My friend and fellow book-clubber Chieko Murasugi came up with the title "Speaking of Genetics." I am grateful to members of my laboratory who have endured my distraction with this project and the Howard Hughes Medical Institute and Department of Medicine at UCSF for their financial patronage during this period. I acknowledge my sister Mary and daughter Annie for their unflagging love and support and, above all, my parents Len and Jean Gitschier, who instilled in me the precious gift of life-long curiosity.

It is a privilege to sit quietly and give full attention to someone's story, and I have been enriched immeasurably by this process. To those of you with whom I have shared the intimate hour or two discussing your life and who trusted me to tell your story, I am deeply grateful.

In the Tradition of Science

An Interview with Victor Ambros

The word "gene" typically brings to mind a stretch of DNA that encodes a protein through an mRNA intermediary, but many genes render a different type of functional information in the form of "non-coding" RNAs. One very important class of non-coding RNAs is that of "microRNAs," extremely short RNA molecules that can interact with mRNAs, thereby triggering a cascade of activity that either destroys the mRNA or arrests its translation into protein. This impeding process, called "RNA interference," was understood and utilized experimentally even before microRNAs were discovered. In this interview, we talk about how microRNA genes were first observed (in the worm C. elegans) and how the connection to RNA interference was recognized.

Interviewed September 30, 2009

Published March 5, 2010

I WAS ON SHAKY FOOTING WITH RNA INTERFERENCE (RNAi) and microRNAs (miRNAs), and I knew I had to do something about it. As the number of miRNAs in humans escalated and I tried to sort through the twists and turns of the compelling story of their discovery, I turned to a colleague for insight. "Interview Victor Ambros," he said, and I took his advice.

For those of you who might also benefit from a little primer on the topic, RNAi is a well-established phenomenon of using double-stranding RNA to effect gene silencing, and it flourished as an investigative tool years before its connection to the tiny endogenous miRNAs was made. RNAi had been first recognized in plants as a response to infections, and the cellular machinery, such as Argonaut and Dicer, to effect RNAi had also emerged. But these advances had been made without appreciating the cellular fleet of stealth molecules—miRNAs—that had piloted under our radar, scanning and tempering our genome.

I asked Victor Ambros to fill me in on some of these discoveries, moments he shared with his wife Rosalind and his long-term scientific collaborator and friend, Gary Ruvkun. After I had to abort plans to visit Victor in Massachusetts, we eventually settled on a Skype interview, and I persuaded him to shoot his own photo on his computer's photo booth. We had a grainy connection but a lot of fun.

Victor grew up on a small farm in Vermont, his father and mother having made the commitment to a rural life, where they set about raising a family of eight children. He went to MIT for undergrad, grad, and post-doctoral work, ventured down Massachusetts Avenue to Harvard for his first job, and managed to slip out of state to Dartmouth for his second. He then returned to the Boston area where he has now settled in at the University of Massachusetts, Worcester.

Gitschier: What do you think funneled you into a career in science?

Ambros: I'm not sure. My earliest recollection was that I dreamed of being a baseball player. But that was until about age 8 or 9. After that, I can't recall not wanting to be a scientist, and I must trace it to reading books that were lying around the house.

I just got intrigued by the tradition of doing science. I read a book about famous inventors and books about astronomers, and decided I wanted to be an astronomer. These were plans and dreams that just sort of came together without any kind of authentic, realistic experience. Just a child reading books and deciding that's what he wanted to do. It seemed like it was a wonderful tradition to be part of—that tradition of scientists and inventors.

Somebody got me a toy telescope when I was young, and I became an amateur astronomer when I was 11 or 12. I built a telescope out of a book. My father encouraged me an awful lot. He was excited that I was interested in science and he would help me with building projects.

Gitschier: He was a hands-on kind of a guy.

Ambros: Yeah, my dad is exceedingly clever. I'd say he is a brilliant man who, because he was born at the wrong time in Europe—in Poland—was caught up in World War II. He went to high school only for a year or so because the schools closed down at the onset of the war. He became essentially a fugitive from the Russians and Germans in Poland. He was captured by the Germans and spent the rest of the War as a forced laborer. He spent from the age of 15 to 19 having no education at all.

When he was liberated by the American army, he worked for the army as an aide to some army officers, and he was exposed to a lot of books in the mansions of the ex-German rich folk, which were being used by the American army as headquarters. That's how he began to teach himself English.

By the time I was born, in 1953, I came to know my dad as someone who was very, very clever, could build almost anything, and was very well-read. It was fun to listen to him talk about books that he had read, and even today we recommend books to each other and discuss them. He speaks four or five languages. My dad is someone whom I admire enormously, especially because I felt that he was someone who had missed an opportunity to be a formally educated person, but he still made a great life for himself and his family.

I remember from a very young age being very conscious of pleasing my dad because of the contrast between what I felt I had, which were all sorts of opportunities, and the opportunities that he missed. So that would help keep me on track—study hard, because after all, that'll please Dad. So he was a very important person for me throughout my childhood and high school. He still is!

Gitschier: Do you mind if I just follow up a little bit more? When you said he was in a forced labor camp, do you mean he was in one of the concentration camps?

Ambros: He's not Jewish; he was Catholic, so he was lucky enough not to be categorically sent to death. He was also able-bodied, so he became incorporated into this system of forced labor that they had in Germany. It was very much like American slavery. People were property. They were essentially rented out or leased to others who were doing work for the government. So my dad became property of the German government, and he worked for a company that processed wood into fuel for trucks.

Gitschier: What happened to his siblings and his parents?

Ambros: Well, his mom and dad had already died when he was still a child. But he had a sister. He lost track of her during the war, but they were reunited in 1960. The Red Cross had a system of registering displaced persons. Eventually names were matched up but it took some years—this was pre-computers. They did not know that the other had survived the war.

She came over here and lived near us until she died just last year.

Gitschier: Do you think you've stayed in New England all these years because of the proximity to your family?

Ambros: Yeah, I would say so. I like New England, and it is nice to be within striking distance. But I did go to MIT, not because it is in New England, but because it was the place I wanted to go to school. We ended up being dug in, in Boston. Also, I'm a person who does like the familiar. Given a choice, I would stay.

Gitschier: Let's move on to some of these major discoveries that you've made. And let's start with your being in Horvitz's lab and working on these things called heterochronic mutants.

Ambros: The term heterochrony referred to a mode of developmental change in evolution, where animals would acquire some change in the relative timing of events, and that would lead to changes in morphology. The classic example is the axolotl, in which the adults retain their gills instead of going through the metamorphosis. Stephen Jay Gould had written extensively about these in his columns in the *Natural History* magazine. When Bob and I started studying the mutants that had primarily changes in the relative timing of events, we thought it would be cool to co-opt that term to describe the mutants, since the term was already there.

Bob had set up this group at MIT that was bringing a really interesting approach, I thought, to this worm, which was to isolate mutants that were defective in egg laying. And Bob's brilliant insight was that there are so many different ways that a worm could be defective in this behavior of egg laying that it allows access to all kinds of processes and pathways in the animal. A worm can fail to lay eggs because it's missing the apparatus and those would include all sorts of developmental mutants, and from that came the heterochronic mutants, which are the developmental timing mutants that lead to morphological problems in egg laying, and all the signaling mutants, the Ras pathway.

Gitschier: But he couldn't have known at the time that there was going to be a Ras-pathway mutant.

Ambros: It's hard to know what Bob actually anticipated. I think that he anticipated more than we give him credit for—whether it was Ras or FGF [fibroblast growth factor] or you name it, he knew that the animal was developing with enormous precision. Cells were talking to each other and neurons were connecting with muscles. So he got mutants in muscles, nerves, neurotransmitters, development, cell lineages, etc. So actually, these heterochronic mutants were a small subset of a whole series of different classes of mutants that were coming out of those egg-laying screens.

So he assigned the project to me to look at the first of these, which was *lin-4*, and another gene called *unc-86*. But I didn't really get any traction with *unc-86*.

lin-4 was the gene that I actually made some progress on, and that was because suppressors of *lin-4* arose spontaneously. One of the first was isolated by Chip Ferguson, who was in the lab at the time. Chip gave this mutant to me and said this mutant suppresses *lin-4*, and it turned out to be a mutation in *lin-14*.

That made the link between *lin-4* and *lin-14*, and my contribution was to find some dominant mutations in *lin-14* that had the same phenotype as *lin-4*. So, the loss of function in *lin-4* was equivalent in phenotype to a gain-of-function mutation in *lin-14*. And then we did some epistasis work and decided that a parsimonious scenario was that *lin-4* repressed *lin-14*.

Then Gary Ruvkun came to Bob's lab. He was a molecular biologist, and nobody in the lab was doing molecular biology. So Gary taught us how to make DNA and do restriction digests. Gary and I collaborated on trying to clone *lin-14*. We made some progress, and we eventually published a paper showing that we had cloned *lin-14*, without including a sequence! In those days you could get a publication by demonstrating that you had identified a band on a Southern and a piece of cloned probe that represented the gene.

Then Gary focused on the *lin-14* project in his lab at MGH [Massachusetts General Hospital] and I took the *lin-4* project to my lab at Harvard.

Gitschier: I know that you and Gary are very close. Was it part of the design that you were going to stay physically close together in the Boston area?

Ambros: No, that was just accidental. And splitting up the genes was a good idea. In those days, cloning a gene wasn't that straightforward. You didn't have a genome sequence. We were cloning genes purely based on mutation. Transformation rescue hadn't been established yet. Each of us started in our labs in '84...

Gitschier: It was almost a decade then before . . . [you published on *lin-4*]. . .

Ambros: [Laughter] Yeah. Well, we had lots of other projects. What was done in my lab was driven by the interests of the students and the post-docs.

Also, *lin-4* was a tough project because there was only one mutation. Even though Bob had been screening and screening for egg-laying defective mutants, and *lin-4* was an egg-laying defective mutant, there was only one allele!

Gitschier: In retrospect, do you think that told you that it was going to be a really small gene?

Ambros: Well, we had lots of concerns. There were categories of concerns. One would be that it was a peculiar mutation, called E912. E means it was identified in England in [Sydney] Brenner's lab, it was actually induced by ^{32}P degeneration. John Sulston had been making ^{32}P-labeled worms for doing Cot curves, and a member of the lab screened the progeny of those animals for mutations and ended up getting E912.

So, we were concerned that maybe it's a peculiar kind of [DNA] rearrangement that fuses this thing to that thing in some way and has what's called "neomorphic" activity. So you might be cloning a locus that in retrospect might not really tell you anything about the normal function of either of the respective genes.

Gitschier: Did that kind of concern make it a back-burner project?

Ambros: Exactly. To clone these genes, we like to proceed by getting multiple mutations, so that when we get to the gene, we'll be able to identify these mutations.

It really wasn't until my wife, Rosalind Lee, joined the lab, which I think was in 1987, that this really seemed like a perfect project for a research assistant. She was a technician and her career didn't depend on this. She came to try to move along the genetic experiments we were doing to find the locus.

And then Rhonda Feinbaum joined the lab as a post-doc. Rhonda was interested in the project, especially as this would be a team effort between her and Rosalind; it wouldn't be all on one person's shoulders. And over the course of 4 years, the two of them ended up, by dividing up the effort, putting together that story which we ended up publishing in 1993.

Rosalind [known to her friends and colleagues as Candy] did the positional cloning, mapping *lin-4* with respect to recombinant chromosomes between two worm strains, and eventually found the DNA lesion associated with the *lin-4* mutation using Southern blotting. And it turned out it, the lesion proved to be a deletion and rearrangement, consistent with a ^{32}P degeneration.

Rhonda did complementation rescue experiments, and she and Candy together made the constructs and whittled the gene down. At some point we recognized that the gene product couldn't be a conventional protein-coding gene because we narrowed it down to what looked like an intron of a protein-coding gene. We were able to get rescue of the mutant phenotype fully with just the little piece of DNA—about 700 base pairs.

In parallel, Rhonda also did a very massive screen for new alleles and picked up one new allele, and that turned out to be really useful. This was a point mutation affecting a single nucleotide of the miRNA. It was the lynchpin supporting the conclusion that that the small RNA was the lin-4 gene product.

Gitschier: By the time you figured this all out, it was what, 1991 or '92?

Ambros: Yes, more or less. We wrote this down somewhere! We knew the basic story for almost a year before we published. There was a year or so where our lab and Gary Ruvkun's lab were cleaning up loose ends and putting together a nice pair of complementary papers.

The importance of that relationship with Gary's lab was that Gary was pursuing an analysis of gain-of-function mutations in *lin-14*. Those dominant mutations that were causing *lin-14* to be essentially constitutively expressed developmentally and de-repressed from *lin-4* activity were in the 3' UTR.

Gary had cloned a gene from another nematode species and did RNA alignments to try to identify conserved sequences that may be important. He was developing a hypothesis that the 3' UTR was a site where the lin-4 gene product would bind. So it became really important to find out what the lin-4 gene product was.

We anticipated that the stories would converge, so we were staying in touch. When Rhonda and Rosalind were zeroing in on this small piece of DNA and showing ultimately that the transcript from that region was really short—the main transcript was only 21 or 22 nucleotides. At that point we shared the sequence with Gary because he said, "We have sequences from these two species, and we should line them up and see whether there is some sort of anti-sense base pairing."

It was actually pretty obvious once we did the alignment that there had to be anti-sense base pairing. *lin-4* matched the *lin-14* 3' UTR in several places, and all of those places were conserved between the two species. And Candy had shown that *lin-4* was conserved between those species, as well, so here we had a little, well-conserved RNA, and the complementarity to *lin-14* was conserved. That was really cool.

Gitschier: Were you feeling pressure from Gary to get this *lin-4*?

Ambros: I'm sure there was a healthy competitive component there. I was thinking, "Well this matters to somebody else, so we really need to push it forward."

Gitschier: Sounds more collaborative, though, than competitive.

Ambros: Well, from my perspective the competitive aspect was [that] you wanted to do at least a good a job as Gary was doing in his part. The pressure was: you don't want to get your part wrong! It was very nice that the sharing of the data and looking at the RNA sequence together came from a desire to make sure that the experience was a good one—good for me and for him. I didn't want to be the one who missed it when he got it. And I didn't want to be the one who got it if he missed it, because that wouldn't feel right either. So we said, we'll exchange sequences and we'll look at it together.

Gitschier: Were you on the phone while you were both looking at it?

Ambros: We sent the sequences to each other and said let's look at this and call back this evening and see what we see. So we called and said, "Do you see it?" "Yes, I see it!"

Gitschier: That must have been really exciting.

Ambros: It was. And it felt good. That we had found something and that we had found it together.

Gitschier: OK, now let's talk briefly about RNAi and how you eventually realized that *lin-4* fitted into that story.

Ambros: The phenomenon of RNAi had been described and studied in plants before Fire and Mello had hit on this double-stranded RNA [dsRNA] trigger concept in 1999. So it was already known that this phenomenon in plants and animals seemed to smell the same—an epigenetic gene silencing mechanism.

That didn't immediately help us appreciate miRNAs. We had found *lin-4* in 1993, and even though we showed it formed a dsRNA precursor, we didn't connect that to the dsRNA-based phenomenon called RNAi that Fire and Mello found.

Gitschier: Because at that time, the RNAi people were talking about things that were brought in to the cell.

Ambros: That's right. At that point, it was a mysterious capacity for the animal to respond. So it wasn't clear what this represented in terms of endogenous mechanisms. The animal was too good at it for it not to be deeply important. And then there was a rapid series of discoveries, where Mello, in one of his important contributions that helped win the [Nobel] Prize for him, was finding Argonaut—that there were these conserved proteins that were required for the silencing in worms.

But, it wasn't until David Baulcombe found that the silencing process in plants involved the formation of very short—about 22–25 nucleotides—dsRNAs, indicating that in plants, and probably by implication in animals, the long double-stranded RNA precursor was being processed to a short molecule.

And I remember seeing that result and thinking, "Hmm, that looks a lot like *lin-4*."

The point of the story I'm going to tell you now is how interesting it is that we—at least I—was so resistant to a new idea. We thought that *lin-4* could be just specific to worms, because Candy had actually tried to find *lin-4* molecules in other species and couldn't find them. And now we know that this little RNA isn't conserved well enough to detect by hybridization.

So, Baulcombe has found this stuff in plants and it's associated with RNAi. So maybe this helps explain how *lin-4* biogenesis works; it must be that it has co-opted the RNAi machinery to be processed. So, you see what I'm saying—not that *lin-4* represented some broad class of things...

Gitschier: ...that were fundamental.

Ambros: Yeah. Maybe it's just a special case of how a system can co-opt the RNAi machinery to make a gene product that is a small RNA product.

And it wasn't until finally Gary found *let-7* in *C. elegans*, with a sequence completely different from *lin-4*, that [we realized] in nematodes the same thing had evolved twice. But personally, it didn't trigger me to think that *lin-4* and *let-7* must be part of something very broad and conserved in all animals. Still, in my mind it was a special nematode thing.

Until Gary published his *Nature* paper in 2000 showing that *let-7* was conserved in sequence in all these animals—sea urchins, mammals... And that was a total revelation. It was a watershed discovery that made me instantly go from a pessimist to an optimist. I said to myself that there must be more small RNAs like *lin-4* and *let-7*, and in other animals. It was really exciting. To open that *Nature* and say, "Holy cow! The way I've been looking at things is totally wrong."

Gitschier: Didn't you know about Gary's work before the paper came out?

Ambros: He kept it secret! It was really cool.

Gitschier: I loved what you said in the Lasker Award commentary [in *Nature Medicine*] that after reading their paper you literally had to sit down for 10 minutes and look out the window to reorder your view of the universe.

Ambros: Right, so Candy and I immediately started cloning those things, and we were very naïve; we thought we were the only ones doing it. That was another adventure.

Gitschier: Actually, that was another thing I liked about both your and Gary's commentaries. It was just so great the way you referred to your spouses as being important contributors to the work. I was touched.

Ambros: I think that what Gary and I were trying to get at, independently, was the question of what's the point, really, of an award, like the Lasker Award? Basically, I feel lucky to be there, because if it weren't for a whole lot of stuff that I did not control, the award wouldn't have happened. If I hadn't happened to work in David Baltimore's lab, I probably wouldn't have been noticed by Bob Horvitz, and if I hadn't been in Bob Horvitz's lab, I wouldn't have even worked on this system. And, if Candy hadn't come to work in the lab, none of this project would have happened. And, if my father hadn't encouraged me... It just gets out of control if you think about what can lead to a moment like getting an award.

And so it has very little to do, frankly, with the particular person getting the award. What the award represents is a process that involves interactions amongst many, many people. And the end, one person ends up getting the award. It's really important to try to acknowledge that and understand the fact that really everything that happens in science, including the discoveries that people try to acknowledge by awards, are really the products of this confluence of people's histories and people's interactions. I really believe that science gets done by people with average abilities and talents, for the most part, and when something special happens, enough so that people want to acknowledge it with an award, it was really... in large part... luck!

We try to say to the public, here's an award for somebody who's really, really special. But actually, it's not the *somebody* who is really special, it's the *science* that is special. The way we do science, and the way it works is so amazing. I wish non-scientists would better understand this. That science is a community exercise, that it involves people interacting, that it involves a lot of good fortune in the context of people trying to do something really carefully and following curiosity. That's why it works so well!

You're preaching to the choir, but the idea is that science remains fun and it is a tremendous adventure. It's great that you do these columns because it reminds people of why we all do science. It's through the stories that we are reminded.

On the Track of DNA Methylation

An Interview with Adrian Bird

While guanine, adenine, thymine, and cytosine form the canonical quartet of bases in DNA, 5-methylcytosine is an abundant constituent of the vertebrate genome. This base, which is generated as a reversible modification of cytosine within DNA, plays a vital role in the "epigenome," a configuration of modified DNA and DNA-binding proteins that can influence gene expression during development or disease. This interview investigates how pockets of unmethylated DNA, known as "CpG islands," were discovered to be associated with active genes and how defects in a protein that binds methylated DNA causes a devastating neurological disease called Rett Syndrome.

Interviewed June 16, 2009

Published October 16, 2009

THE DAY SCHOOL LET OUT FOR THE SUMMER, my daughter and I packed our bags for Britain, where we had lived for a few months in 2006. Annie was eager to reconnect with her friends there, and I had arranged to conduct three interviews. In desperation and with the clock ticking, I struggled to fit my bulky recorder into my wheelie when it dawned on me that the "talk app" on my daughter's iPhone should be up to the job. You can imagine the reluctance and skepticism on the part of my 15-year-old, but she managed to get into the spirit and acquiesced.

First up on my schedule was Adrian Bird, who holds the Buchanan Chair of Genetics at the University of Edinburgh and is also Director of the Wellcome Trust Centre for Cell Biology. Long before the word "epigenome" was coined, Bird began mapping the distribution of DNA methylation (occurring at the cytosine of CpG dinucleotides) in the genomes of a variety of species. His work emerged just as agarose gels, restriction enzymes, and Southern blots were being developed. Bird later spawned the idea of CpG islands, pockets of DNA rich in unmethylated CpGs and frequently found in conjunction with the promoter regions of mammalian genes. Bird's observation provided a roadmap for disease gene discovery for about 15 years, until human genome draft sequences began to emerge.

Bird's laboratory then went on to identify proteins that bound to methylated DNA, one of which (MeCP2) was discovered years later to be defective in Rett Syndrome, a rare X-linked disorder in which affected girls develop autism and a distinctive set of behaviors. This astonishing turn of events propelled Bird to extend his studies on MeCP2 to a murine model for Rett Syndrome, ushering in new ideas about therapy for this devastating illness, but still leaving open the question of MeCP2's role in the brain.

Bird and his wife Cathy Abbott, also a geneticist, invited me to spend the night prior to the interview with them (future interviewees, take note!), and I was delighted to do so. Still jet-lagged, I traveled by train, leaving behind the uncharacteristic sun of Cambridge to find cold rain penetrating the skylights at Edinburgh's Waverley Station. It felt very cozy to share the evening with them and their children, Tom and Annie: chatting, watching some Twenty20 (an abbreviated form of cricket), playing Uno, feeding the three guinea pigs, and experimenting with the iPhone's tape app, which, to our delight, worked.

Gitschier: My first question is a two-part, integrated one. How did you get interested in methylation, and what was the state of the art at the time you started working on it?

Bird: I first got interested when I was in Zurich doing a post-doc.

Gitschier: Whom were you with there?

Bird: Max Birnstiel. I had been in the States doing a post-doc [with Joe Gall at Yale] on gene amplification in frog oocytes. When I went to Max's in Zurich, he had a visitor named Ham Smith.

Gitschier: What year approximately was this?

Bird: 1973–1974. Ham was on sabbatical, and the first thing he did was to make a restriction enzyme, HpaII—*Haemophilus parainfluenzae* II.

I was making ribosomal RNA genes, just for something to do really. We knew there was a difference between the amplified ribosomal RNA genes, which were extrachromosomal in the oocyte, and the chromosomal ones, and that the difference was due to methylation. Don Brown and Igor Dawid had shown that chromosomal rDNA had 5-methylcytosine and the amplified didn't.

So I used HpaII to cut this amplified [extrachromosomal] DNA, and it cut beautifully. But when I tried with purified chromosomal ribosomal DNA, it didn't cut. It was known that restriction enzymes were blocked by methylation in the organisms from which they are derived; there is usually a restriction enzyme and a modification enzyme that matches it and that protects the genomic DNA of the host from its own destructive enzymes. So it seemed that some of the methylation [in the chromosomal rDNA] might be mimicking the blockage that occurs in the *Haemophilus parainfluenzae* endogenous enzyme.

Gitschier: Did Ham Smith know that HpaII didn't cut methylated DNA before you did the experiment?

Bird: Probably, but he just made restriction enzymes in order to make himself at home. Restriction enzyme technology was new, and agarose gels had just been brought in. And another guy there was Ed Southern, who had just invented Southern blotting, so there was a bit of coincidence here.

Gitschier: Was Ed on sabbatical?

Bird: Yes, he was as well. I can't remember whether they were there at exactly the same time. I think they overlapped.

Gitschier: So, why is everybody coming to Max Birnstiel's lab?

Bird: Max Birnstiel and Don Brown were hotly competing groups because both of them worked on ribosomal RNA genes, which you could purify by buoyant density centrifugation. Before you could clone DNA, the only way to get hold of pure gene was A) because it was highly repeated and B) because its buoyant density was different from the bulk [of the DNA]. So if you ran enough cesium chloride gradients you could get it pure and then you could study the structure of it. Frog ribosomal genes and sea urchin histone genes were where it was *at* in those days.

Gitschier: What attracted Ham to that lab?

Bird: Max and Ham knew each other and were friends.

Gitschier: So for him, it was going to a friendly and interesting place. And Ed?

Bird: He came over to work with Max, I have to say, partly because he was going out with someone who happened to work in Max's lab. In those days, people's motives were more random than they might be made to look now.

Gitschier: Sounds lovely. So Ham's making this enzyme, and you are just trying this enzyme, you don't know what the results are going to be.

Bird: I'm not sure what I'm doing, to be honest! I knew I wanted to do something interesting, but I was just playing around more than anything else. Ham was playing around with the enzymes he knew about and Ed was doing the technological things that he really liked doing. So accidentally this gave rise to the idea that one could use restriction enzymes to map methylation in DNA. There was a conjunction of areas that were needed before one could exploit this properly, and that didn't happen until I got back to Edinburgh.

By 1975, I was back there in an MRC unit and mapping the methylated and nonmethylated sites in the ribosomal genes in *Xenopus laevis*. And it took absolutely ages to get that published.

Gitschier: At some point you moved away from frog oocytes.

Bird: We looked at sea urchins—invertebrates—and found there was both a methylated and an unmethylated fraction of the genome. The last things we came to were vertebrate cells and they didn't seem to have anything like what you see in the invertebrates. When you digested the DNA with these methylation-sensitive enzymes, nothing happened because most of the DNA is methylated.

Then we had the idea that maybe we could see a small fraction of unmethylated DNA if we end-labeled it. That was the work of David Cooper who did a PhD in my lab. I can remember him doing the first end-labeling, because it made a horrible blob. Something that could be artifactual, but it wasn't, and we spent quite a lot of time showing that. I cloned mouse fragments derived from that blob, and that was our 1985 paper in *Cell*. And then we restriction-mapped them and showed that they came from clusters of nonmethylated CpGs in the genome.

We were not totally alone in reaching these conclusions. Tykocinski and Max had looked at DNA sequence in MHC class I genes and saw these clusters, and earlier a guy called De Crombrugghe also saw something like this. There was also stuff on the inactive X chromosome—Barbara Migeon had seen non-methylated sequences at the 5′ ends of genes.

It was in the air, but not yet an accepted generalization. Our data really suggested that there was a category of genomic DNA that was full of CpGs that weren't methylated and that eventually were understood to be near promoters. So we kind of brought it all together.

Gitschier: So you published this review article [in *Nature*], which is where I became familiar with your work.

Bird: Yes, that was cited loads of times, because it was the mapping phase of the genome project where people wanted to map themselves into reality.

Gitschier: Well, people used them to find genes. They were like little flags that said, "Hey over here, I'm a gene!"

Bird: Exactly.

Gitschier: Did you coin the name "CpG islands"?

Bird: No I didn't. We called them "HTF islands" for "HpaII tiny fragments", but reviewers said, "What the hell is HTF?" It was Marianne Frommer, who was doing a sabbatical in my lab at the time the *Cell* paper was published, who called them CpG islands—it made more sense.

Gitschier: OK, let's switch gears. At some point, you started to work on proteins that bound to methylated regions.

Bird: That arose by chance as well. The CpG islands provide an approximate way to map methylation throughout the genome, but at the time there was no way to do that properly. So we kind of ran into a brick wall—what you really want to know is where the methyl groups are throughout the genome.

Gitschier: And you found out where they weren't.

Bird: So I decided to work on how the CpG islands might originate. We asked, "Does something bind to the nonmethylated sequence that might protect it from methylation?" Just by steric inhibition. We made an oligonucleotide, and at the time it took Amersham about 4 months to do it. I just made up a sequence full of CpGs that were in restriction enzyme sites so we could test their methylation status, and then we oligomerized them and methylated the sites using commercial enzymes, because you could buy HhaI and HpaII methyltransferases.

Then we made extracts from mouse liver nuclei. And what we found was something that bound to the methylated one but not the unmethylated one. So after a period of time where we considered whether this was interesting or not, we decided to work on isolating what bound to the methylated DNA.

Gitschier: What year are we now?

Bird: This is 1984–1985.

Gitschier: So, if you already knew about this protein you called MeCP1, why did you go on to look for more methylated-DNA binding proteins?

Bird: We were trying to purify MeCP1, and we were mucking about changing the assay, and in doing that we detected another protein. We couldn't purify MeCP1 for love nor money, but it has been purified subsequently by Yi Zhang. However, we could purify a different protein, which we called MeCP2. It was relatively easy.

Gitschier: What, then, did you do with ... MePC2 [sic]?

Bird: Yeah—terrible name. Entirely my fault. Like HTF islands. I have a talent for inventing awkward names.

First thing we did was dissect out the methyl-CpG binding domain. These were not high-profile publications, because no one was interested in MeCP2 for a long time. Then we wanted to know if it was a methyl-binding domain in vivo, as opposed to in vitro. And that is STILL an issue that people are not convinced about. But I am.

We showed that it went to the satellite foci in mouse nuclei. In mouse, the pericentromeric heterochromatin is quite CpG rich and heavily methylated. MeCP2 goes there and stains in "spots". It's about 10% of the genome, but contains half of all the methylation. It was a nice visual way of showing binding. Then we used a DNA methyltransferase mutant that has much lower levels of methylation, and we no longer had staining of the heterochromatin. So that said that it was the methylation that was causing it to bind. That was quite an important paper because it showed that this mindless, in vitro assay that we used for purification actually had some biological relevance.

Next we decided we wanted to know what MeCP2 did, so we knocked it out. And then Zoghbi and Francke showed that it was mutated in Rett Syndrome.

Gitschier: I didn't appreciate that. You actually did make a knockout before they found it as the basis for Rett?

Bird: Yes. The reason you didn't appreciate it was because we didn't get the right result. In something like 1993, we knocked it out. Because it's on the

X chromosome, you've got a null immediately in male ES cells. We made chimeras and the chimeras all died. We concluded that it was an embryonic lethal.

By the time the Rett Syndrome story came out, we had already decided to do it again and made a conditional knockout. I now think that growing ES cells in the absence of MeCP2 somehow compromises their ability to form embryos properly.

Gitschier: So you didn't get a result that was really wrong, you just didn't get a result that was really useful in studying Rett Syndrome.

Bird: Correct. We actually started on this conditional MeCP2 knockout before the Rett Syndrome story came out. We knocked MeCP2 out in early embryonic development. Those [male] mice were born and normal until about 6 weeks and died at about 12 weeks. And then we did it again just knocking out in neurons and glia with the same result. This said the phenotype was entirely due to the brain.

Gitschier: OK, so you're working away on trying to understand the function in mice, and suddenly Zoghbi and Francke labs discover that the gene you've been working on has real human consequence.

Bird: Yeah.

Gitschier: Were you a reviewer on that paper?

Bird: Yes.

Gitschier: What did you think when you saw the paper?

Bird: I thought, "What the hell is Rett Syndrome?"

Gitschier: I was going to ask you that!

Bird: In my collection I did have papers about it, but I hadn't been keeping up with it at all, honestly.

Gitschier: What else did you think?

Bird: I thought, well we're already working on these mice, so now let's see if we can make a model for Rett Syndrome as well.

Gitschier: But what about the physiological or mechanistic implications? The fact that defects in a methyl-CpG binding protein cause a neurological disorder seems wild to me.

Bird: We're working on that quite a lot at the moment.

I think the people who were looking in the interval for the gene that was mutated were quite disappointed to find that it wasn't one of the GABA receptors, which is in that region, but rather that it was this rinky-dink little housekeeping gene. No one could get terribly inspired by finding it. But to me, it was great.

It is quite exhilarating to realize you have some preliminary insights into something whose function turns out to be very important. I had previously been rather disparaging about medically relevant research, considering that I was working on pure knowledge and that biological information was intrinsically important. But as soon as you collide with biomedical relevance, it changes your perspective. It breathes life into the project and it has added new dimensions to my research. I'm absolutely delighted it happened.

But even now, not everybody agrees about what MeCP2 does.

Gitschier: Let's turn to that now, because to me that is very murky.

Bird: The initial experiments said [the following]: A) It's a methylated-DNA binding protein. B) It's very abundant in brain. In fact, that's where we purified it from. C) It's associated with a co-repressor called Sin3a and is able to repress transcription in model systems by co-transfection in cultured cells. So the prevailing view is that it's a methyl-CpG binding repressor.

But, if you look in the knockout you find a lot of genes go up and a lot of genes go down, whereas you'd expect, if it's a repressor, that genes would only go up. So this has made everyone think twice.

But another argument, which people don't usually find persuasive I have to say, is that some genes going up and some genes going down makes sense if MeCP2 is regulating very many genes. Because you've probably got a closed system—for everything that goes up there must be something that goes down. A good example of this—if you treat cells with the histone deacetylase inhibitor TSA (trichostatin A), this causes histone acetylation to go through the roof, which is a marker of activity, but expression-wise as many genes go up as go down. Everything would LIKE to go up, but you can't employ more polymerases, etc., than are there.

We've been through phases where we have been prepared to throw out all the old stuff and say neurons are doing something different with MeCP2. But the data have forced us to come back to the original idea that MeCP2 coats the genome and recruits enzymes that deacetylate and keep the acetylation low.

So what is its function? Mike Greenberg [Harvard Children's Hospital] showed that MeCP2 is a phosphoprotein. It gets phosphorylated when neurons are active. We know that when neurons fire, you get bursts of synaptic protein synthesis and nuclear protein synthesis and this is associated with plasticity. It is an attractive idea that MeCP2 has something to do with that. Neurons fire—phosphorylation of MeCP2 changes its properties—let's say it comes off, leading to a burst of transcriptional activity, which is somehow involved in neuronal homeostasis. It's an intriguing possibility.

To be honest, the reason that there are so many theories about what MeCP2 does is that none of them has been nailed down experimentally. There is a bit of a vacuum there, and people fill vacuums with speculations, as they should. All of the things I'm saying are subject to different views from different people, but I would say MeCP2 is involved in maintenance.

Gitschier: When you say maintenance—do you mean maintaining the neuron itself, or maintaining the neural activity?

Bird: Functional integrity. What seems to degrade in Rett Syndrome is neuronal activity. The neurons aren't dead. What you'd like to say is that neurons are degenerate, but you're not allowed to say that, because neurodegenerative disorders, like Alzheimer or Parkinson, involve neuronal death.

Gitschier: We think of autism as a neurodevelopmental defect and Rett as being in that category as well, because its onset is during the development of a human being.

Bird: The reason that people are so wedded to the idea of Rett being neurodevelopmental is that of course there are lots of things going on in neurodevelopment at the time girls get the symptoms. What impresses me is that female mice get equivalent symptoms between 4 and 12 months and humans between 6 and 18 months—in REAL time, that is the same, more or less. But developmentally, they are at totally different stages.

People see only these two categories [neurodegenerative and neurodevelopmental], but I would say it's a neuromaintenance disorder. The best evidence for that is reversibility [see below].

Neurons are long-lived complex cells that expend a lot of effort deciding who they are going to be connected to. As a neuron you'd better remember all that because lots of things are going to happen to this organism over a period of years, or decades in our case, and that neuron will not get a chance to renew. So maintenance becomes an extremely important problem, and I think MeCP2 is one of the proteins that have evolved to ensure that.

In my opinion the reversibility experiments that we did are interesting for all sorts of disorders in which neuronal death has not been established, and that includes autism and schizophrenia, for example. The general assumption is that once you've got one of these neurological disorders, that's it. When one sees Down's Syndrome, or mental retardation of any kind, it is ingrained in us to believe that nothing can ever be done to reverse that.

Gitschier: Tell me about the reversibility experiments.

Bird: That's probably one of the most impressive things we've ever done. I'll show you a movie. Neurons don't die in Rett Syndrome or in the mouse model. So this raised the possibility that one could put the gene back and find out whether the symptoms are reversed. Jacky Guy took MeCP2 and inserted a stop cassette in an intron—designed to prevent expression, but flanked by loxP sites. When you mate that with a mouse with Cre expressed under estrogen receptor control, you can inject tamoxifen, release Cre, delete the stop, and presto transcription starts again. It was one of those projects where everything worked even though we thought it might never work.

Here's the life cycle of a male MeCP2 stop-cassette mouse. It is born and then gets symptoms at 6–8 weeks and dies at 9–16 weeks. We inject MeCP2 here, when the mouse has advanced symptoms and is near death. It doesn't really move, very low to the ground, feet splayed apart, tremor, arrhythmic breathing. Now, here is the same mouse 4 weeks later.

Gitschier: Oh my goodness. It looks exactly like the control.

Bird: We've done this with many mice and it's consistently reversible. Females breed normally for about 6 months, but after that they hit a brick wall, become inert and obese, develop hind limb clasping, tremors, and arrhythmic breathing—all the things that mimic aspects of Rett Syndrome, and it is stable, just as in humans, and lasts for the rest of their lives. That is the real model for Rett Syndrome. These animals are way beyond neurodevelopment. But even at this late stage, it is reversible. It establishes the principle that Rett Syndrome, as least in mice, is reversible, and it encourages the belief that it might be reversible in humans, too.

I must say the reversal is the most surprising result we've ever had, partly because it went completely against what was expected and partly because I'm a biochemist at heart and this was a sophisticated genetic problem. It was nice to follow the problem into the mouse and do an experiment that has had an impact beyond Rett Syndrome and that many neuroscientists find interesting. I'm now on the fringes of that world, and I'd like to try and make more contributions there.

Gitschier: So you're not about to stop soon, I take it!

Bird: Everybody knows that you yourself are probably not the best person to judge that, but I don't feel in the slightest like stopping.

When people write and say that the mouse reversal has transformed their view of prospects for their daughter, you are flattered by that, but it very quickly gives way to frustration—that you've raised people's hopes but you cannot in any way replicate the reversal in humans. Although it seems like a short step to a lay person, in fact, it is a gigantic leap into the unknown.

Willing to Do the Math

An Interview with David Botstein

In the 1970s, human genes were examined for the first time, and gene sequence variations, also known as "polymorphisms," were discovered. It was soon suspected that the human genome was littered with such polymorphisms. This interview chronicles the genesis of an idea that DNA polymorphisms could serve as genetic markers to map the location of disease genes, thus paving the way for a genetic map of the human genome. It also explores a master teacher's philosophy behind an innovative curriculum.

Interviewed February 22, 2006

Published May 26, 2006

DAVID BOTSTEIN'S NAME CAME UP TWICE when the *PLoS Genetics* editors tossed around ideas for potential interviewees, and with good reason. First, he is now director of an exciting development on the Princeton University campus, the Lewis-Sigler Institute for Integrative Genomics, which weaves the physical, computational, and biological sciences into a cohesive endeavor. Second, his scientific discoveries have run the gamut of organisms, from phage P22 to yeast to humans. And third, David is no shrinking violet; he has a deeply held opinion about everything, and he isn't afraid to voice it.

For those of you who are familiar with David, these words will likely fly off the page. There is little need to ply him with pre-formed questions; once you get him going, he is a verbal tsunami. His zeal is legendary, and his delight in people and ideas is palpable.

I met David at his office in the Carl Icahn Laboratory on a chilly February morning after the snow had melted. The building is the elegant design of Rafael Vinoly. Its airy entranceway, which spans the length of the curved glass façade, embraces a small café, a cylindrical seminar hall, and, jarringly, the 30-year-old lead-clad prototype of a Frank Gehry house, which David aptly refers to as the "armadillo." Double-helical shadows are cast by latticed aluminum pillars that hug the portico's arc, and altogether too stylish, white, Tom Vac and Pierre Paulin "Orange Slice" chairs populate the space. Yet tucked into a corner were six dowdy Revcos and an ice-o-matic, signaling that this building is not just a pretty face.

I powered up my fancy new digital Marantz recorder, stoked with a high-capacity flash card, and pushed the record button. Here follow excerpts from what proved to be the first in a double-header interview.

Gitschier: Let's start with the Lewis-Sigler Institute. What are you trying to do here?

Botstein: Actually, I came here to do something about science education, an experiment, if you like. As you know, I was at MIT [Massachusetts Institute of Technology] for many years and taught undergraduate and graduate courses, and was director of the graduate program. Beth Jones and I invented the whole project lab system. We were instructors at that time [1969], a non-tenure track appointment. You worked in somebody's lab, but you did teaching.

Gitschier: Whose lab were you in?

Botstein: I was with Maury Fox. Beth brought yeast to Boris's [Magasanik] lab. I brought P22 to Fox's lab. We weren't randomly chosen to do this. I think the idea was that they brought people who were particularly good experimentalists to teach these lab courses.

We were given this really awful course to teach, and we made a proposal to teach it in a better way, which involved major changes in the curriculum, and they [the older faculty] went along with it. And in fact, I got a job [as Assistant Professor] out of it.

There is something about teaching that makes you a better researcher. I know this is very countercultural wisdom, but I believed it all along. Luria, Magasanik, and Levinthal all believed it. Levinthal and Luria both had a very strong influence on me in this regard.

That's one part of the story!

The other part of the story is that through the Academy [National Academy of Sciences], I became very interested in the question of why it is that there are fewer and fewer kids in America who are interested in biology. And through the Academy studies, it became clear from the statistics that the total number of kids interested in any kind of science had been falling for the last 30 years. And the conventional wisdom is that it is all about K–12 education.

I don't believe that for a second.

Gitschier: And why is that?

Botstein: I'll tell you the argument. There are three arguments why this is not right.

One is that it is a mistake to think that the high schools were ever so good. There is zero argument to say that they have gotten any worse. If you ask about the absolute number of kids who get calculus in high school, it has risen 10- or 20-fold in the last 30 years. When I went to college in the late '50s, I actually did have advanced placement, but it was rare!

Second argument is that if you look at the outcomes of kids who get degrees in science and then go to graduate school—this is actually a Tom Cech study published in the *Daedalus* just before he became head of the Hughes—and you ask where the graduate students come from, it turns out that small colleges, like Swarthmore, who draw from exactly the same pool as the big universities like Princeton, Harvard, and Stanford, are many-fold better at motivating students to become professional scientists. It's something like 5-fold.

Third argument is that who was teaching whom in the '50s and '60s was very different [than now]. I had biochemistry from Konrad Bloch. This was Konrad Bloch before he got the Nobel Prize. He had nothing better to do than teach 120 kids biochemistry. He did it very well. He took it very seriously. He did not go to the dean and say, "Oh, I have this grant and I'm about to win a Nobel Prize and therefore I should be free to do research."

He understood, as I think we all understood, that there is an organic connection between teaching and research. So I came with all this baggage. And Ira Herskowitz, and basically the MIT school, had that baggage. You remember what it was like! Luria taught 7.01 [introductory biology]; Boris taught microbiology. Gene Brown had 7.05 [biochemistry]. The leaders were leading from the front!

For various reasons, there were issues at Stanford, and I was motivated at least to look at my alternatives. I had been talking on and off with Shirley [Tilghman], because this institute was started and Shirley was going to be the director, and then they made her president [of Princeton]. And she had incorporated the idea, which I think was really very smart, that there was already at Princeton an unusually close relationship between the molecular biologists and the physicists.

There came a moment when I called Shirley and said I had ideas about what could been done with such a place. It was a substantial departure from what she had intended. I submitted myself to the formal process and tried to convince everybody to do what we're doing—which was to say, "Let's see if we can do something for a subset of students who, I believe, are underserved in the university." Students who are not going to medical school, necessarily, but who are actually interested in physics or chemistry or molecular biology—who in high school hadn't advanced so far that they already differentiated into physicists or chemists or biologists.

What happens to students who come to college wanting to learn biochemistry? They find themselves first in a chemistry class with a hundred students with absolutely no interest in chemistry. All of those students drill a hole in the head of the instructors and each other to get the best possible grade because all they want is the grade. You teach these people later, and you realize that they are unteachable, to a first approximation. I have never failed as a teacher, except when trying to teach genetics to medical students.

I knew that reform was impossible. The word reform never passes my lips.

Instead, what I proposed was that we use this building, which could house as many as 15 or 16 new faculty, and use resources in the various fields to mount an alternative introductory curriculum in the first two years—at a very high level for students who are willing to do the math, students who want to learn computation—and offer incoming freshmen a choice. Students could learn in the usual way, with the standard, well-worked out curriculum. Math—you can go back 200 years, and you can't find a new thought. Chemistry is taught in exactly the same way that Linus Pauling prescribed in his book published, I think, in 1926.

Or the students could participate in this new experiment in which we'll teach computer science, physics, chemistry, and molecular biology corresponding to the introductory level in six semesters of work—four in the first year, two in the second year—and then major in any of those areas.

What we did was we had eight senior faculty, a sort of Noah's ark of science, sit every Monday at lunch for two to three hours and simply ask the following question for all of these fields at the most introductory level: Is this problem or this idea or this concept fundamental, or merely traditional? We collected all the fundamentals and did the best we could to make them into a coherent sequence.

We tell the students, "This is not the low-energy path to medical school." When they come here, they have fear and trepidation, but I tell them, privately anyway, "Look, if you don't get too many Cs and you work in somebody's lab and do a good job, you'll get into any graduate school in the country. So stop worrying about the grades. If you get a B in everything that we do, you'll be a heroine or a hero."

The adventurous students, the risk takers, come to us. They know how deadening it is to be part of premedical or preprofessional education. It is a mistake to underestimate the perceptiveness of children, as you know!

The breadth of the program has really been its great strength. Morale is very high. The other day, one of our students came to me and said she was thinking of majoring in computer science. She said, "I knew nothing, and I was terrified of it, and I had no idea. If I hadn't taken this thing, I would never have learned to program, but I love it!" Well, that's what we're here for!

The students had a final exam last spring, worth half their entire grade for the year, and all of them showed up in identical T-shirts they had designed. These students are really terrific. I haven't seen students like this since 20, 25 years ago at MIT. They are really turned on.

We have an endowment for five Lewis-Sigler Fellows. We search the country for the best experimentalists we can find, right out of graduate school, no post-doc. And we say, "Instead of being a post-doc, we'll give four benches, $250K a year for research, and you can stay for five years—but you have to teach the lab part of the undergraduate curriculum." And they are fantastic.

Gitschier: Sounds like you've recreated your old job at MIT! What about the graduate program?

Botstein: The graduate school is a more difficult problem, and it's a problem we haven't solved—yet—but we're working on it. We started this Quantitative and Computational Biology Program. This is being led by Leonid Kruglyak, who graduated from Princeton *summa cum laude* in physics. So the idea is that we are going to do exactly the same thing that we did with the undergraduates, in a way. All the participating departments, of which there are many, are going to admit a bunch of students who might be interested in the other field, and we'll try to teach them a few things together.

We have one such course called Method and Logic in Quantitative Biology, and we teach the classic papers, many of which are forgotten because they can't be taught now because people don't understand the math. You may remember reading Luria and Delbruck, and you may remember that nobody understood Luria and Delbruck because they didn't have the math—and these were MIT graduate students! Poisson distribution—don't bother me.

Gitschier: The teaching enterprise is the big glue that is holding this institute together.

Botstein: And that's part of why they hired me, because they understood that eating lunch together was probably not going to be good enough to actually get people to talk to each other seriously. People need to have some common task that is orthogonal to their own research, and eventually, before you can turn around, they are doing stuff together.

This endeavor is not a solo endeavor of mine. Bill Bialek in physics and in the Lewis-Sigler Institute is the architect of the syllabus itself, and the main lecturer in the first semester. The laboratory parts of the freshman course were designed and run by two fellows—Maitreya Dunham, a geneticist, and Will Ryu, a physicist. These fellows are terrific. I'm organizing everything so that by the time I retire in seven years, everything will run on younger people, and they won't actually need me. I'm no spring chicken!

Gitschier: Let's talk a bit about your roots. Your younger brother Leon is a well-known conductor, and I read somewhere that you also considered a career in music.

Botstein: Briefly. When I was in college, I did take harmony and counterpoint and was in the music scene. I was doing that and physics. But physics was taking most of my time.

What I was interested in was choral conducting. I was a singer. I sang in the Harvard glee club, and I was the rehearsal conductor sometimes. I still remember trying to get people to get the rhythm right in Stravinsky's *Oedipus Rex*. It's not that hard, but it's amateur singers. I don't have such a great voice, but I knew about music.

Gitschier: Are you a baritone?

Botstein: Yes. In those days I could do bass; I had an acceptable F and an audible E, so it was enough. I was pretty good in the midrange. The high notes were not too challenging for me. I sang for many, many years. I sang, memorably, when

Kennedy was killed. A huge pick-up group sang the Mozart *Requiem* in Symphony Hall. That was really an occasion.

Gitschier: Did your parents sing?

Botstein: My mother's family was heavily into music in the old country in Poland. My grandfather was a great patron of the arts; he was a very rich man. There were many famous musicians who he sponsored to go to conservatory from that area of Poland. He survived the Holocaust, and I knew him.

My mother was a pediatrician. She was [Guido] Fanconi's assistant. She was in fact the first to show that cystic fibrosis is inherited.

Gitschier: How did she show that?

Botstein: A family study in Switzerland during the war. Dorothy Anderson did the same thing at Columbia Babies Hospital, and our literature was all Anderson [because Fanconi published only in German]. It was Fanconi, of course, who discovered cystic fibrosis.

Gitschier: I didn't know that.

Botstein: Failure to thrive, just like Fanconi's anemia.

Both my parents were physicians. My father was what you would call today a radiation oncologist. He was the first in this country to use betatrons.

Gitschier: Was your father Swiss?

Botstein: Neither was Swiss. They were there in medical school [in Zurich], and they stayed there as residents and then the war broke out. He was from Poland also. Actually, the family comes from Odessa. The other side comes from Vilnius.

My mother was Fanconi's *oberartz*—the chief resident. Ania Wyszewianska, later Anne Botstein. When she came to this country, she was offered positions in various places, but she started to lose her hearing—she had Meniere's disease. She went into private practice. She worked in one of the first HMO's in New York, a place called the HIP at Montefiore hospital where she was Chief of Pediatrics for 25 years, so she was very successful. My father went academic and was a professor at Einstein. And he is very well known. His residents are all over the country.

Gitschier: So you lived where?

Botstein: In the Bronx, in Riverdale, on the corner of the Hudson River, and on the Yonkers border.

Gitschier: You were born in Switzerland?

Botstein: Yes.

Gitschier: Did you have Swiss citizenship?

Botstein: No, there is no such thing as Swiss citizenship for foreigner Jews. Swiss citizenship is hereditary.

We were stateless, and then my parents got a visa for the United States in 1949. They had applied for an American visa in 1935 when they were married, but the Polish quota was full. Fourteen years later, one of their friends was at the US consulate, and they had just posted a list on the door of who had a visa, and they noticed that my parents were on the list. My parents had two weeks' notice to make up their minds as to whether they would come. And they did.

I remember that trip very well. I was seven. My mother had a cousin in New York, and they had letters of recommendation from their bosses in Switzerland who were very well known. My mother tells the story of taking the boards in pediatrics, and it was all "anonymous." She had Nelson, of the famous textbook in Pediatrics, as her examiner, and they talked about polio and this and that, and then he says, "Well, you really know these statistics very well. I'm not supposed to do this, but I know you are not Fanconi—you must be Botstein!" And then they gossiped the rest of the time. It's the same thing as if Sydney Brenner tried to interview Ira Herskowitz.

My mother is still alive, and she lives in an apartment right next to my brother, and I see them nearly every week now. My brother and I have always been very close. He's the famous one. And deservedly so. He's a real innovator in teaching, too!

Gitschier: Since I'm a human geneticist, I can't help but ask you about the genesis of the 1980 mapping paper with Davis and colleagues [in *American Journal of Human Genetics*].

Botstein: That's famously been written up a number of times. The account in Bishop and Waldholz's book *Genome* is actually very accurate, as I recall.

Gitschier: I want to hear it in your words.

Botstein: I was teaching with [Ron] Davis and [John] Roth, at Cold Spring Harbor, the bacterial genetics course in 1978. Jim [Watson], typically, was too cheap to allow us to get together. Roth was in Utah, I was in Boston, and Davis was at Stanford. So Roth finagled that we, Davis and I, would be outside readers for the training grant that Utah had. The genetics training program had a retreat in Alta.

We had to listen to all the talks, and one of the talks was by one of [Mark] Skolnick's students, the guy who did hemochromatosis, what's his name? If you're gonna make history you gotta look all this up. [Meanwhile, David does a quick PubMed search.] This kid got up and talked about how there were two ways to think about hemochromatosis and one of them involved linkage to HLA [human leukocyte antigen]. And the linkage to HLA would make hemochromatosis recessive. And you could build a reasonable model, and they had a likelihood score for that possibility.

Kerry Kravitz! 1978!

There were, at Utah, a lot of hard heads, mainly immunologists, and these immunologists said, "This is bull; it's got to be physiological because HLA affects all this other stuff, and you guys are talking through your hats." They didn't like Skolnick's approach. They didn't like the whole idea.

Now, of course, it's 1978, nobody knows what a LOD [logarithm of the odds] score is, and the only person in the room who knows, besides Skolnick and Kravitz, is Botstein. So I find myself in the middle of this argument. This argument erupts, and I am defending this student because the student and I understand what he did, and nobody else does, and he has not managed to explain it. Skolnick isn't explaining it either. None of these guys are great explainers. It's a medical school, okay!

So finally I say something like, "Look there is nothing special about HLA. What's good about HLA is that it has many alleles, and because it has many alleles, you can tell if you have linkage, and if you have many multiallelic markers all over the genome, you can map anything!" And as soon as the words were out of my mouth, I look at Davis, Davis looks at me, and we both understand that of course there are such markers, and we could make a map of the human genome tomorrow.

Gitschier: What kind of markers were you thinking of?

Botstein: We were thinking of two things, one of which would now be called SNPs [single nucleotide polymorphisms] and the other, which we expected would be more common, was insertion sequences, movements of transposons—because in yeast, that was what was going on. You see, Davis's part of this was TY1. Remember what my history is. In 1978, what was I known for?

Gitschier: Uh, I give up.

Botstein: Tn10! We were one of the groups that discovered transposons in bacteria. Russell Chan, Nancy Kleckner—making mutations, mobilizing genes. It was a really big deal. So the first thing we thought of was, "Oh, there are going to be transposons all over the place!"

We were surprised to discover that the really good markers were all these CA repeats. We did not anticipate that. But I understood pretty much the whole thing, and then the rest of that meeting was—we sat in the bar, we drank, and we figured out how we could make a map of the human genome: how it would work and how many markers we would need. The whole 1980 paper. Everything in there was there in outline by the end of the day.

I went home and I explained it all to Maury Fox, and he understood why this was interesting, and he understood the math. And I persuaded Ray White to look for polymorphisms because he had been doing this transposon jumping around business in introns and ribosomal DNA in *Drosophila*. And it wasn't going anywhere. He had the right technology. And then he and Arlene Wyman found what is now called D14S1, and we were off.

What was really noticeable at the time was that the human geneticists didn't get it. At all. At all, at all, at all. It took a really long time. Skolnick was beating the drum. In 1983, I went to the ASHG [American Society of Human Genetics] meeting, and I gave this long discussion of how it all would work, and I had to explain Southern blot and this and that. I went to NIH [National Institutes of Health] and tried to get money, and Ruth Kirschstein looked at me and said, "We don't do things like that." That was the origin of her opposition to the genome project, I'm sure.

Then [Jim] Gusella did his number [mapped Huntington disease]. On the one hand, it gave the idea some credibility; on the other, it did everybody a huge disservice, because people started to look with random markers. And the statistics!

Gitschier: They were so lucky; it was the twelfth marker!

Botstein: The eighth! The eighth! The prior probability of finding it in eight markers was zero! I knew that!

Gitschier: Well, for once, ignoring the math paid off!

Wonderful Life

An Interview with Herb Boyer

Recombinant DNA results from a process by which DNA fragments from two different organisms are joined together and made to function as a unit. This process was enabled by the discovery of bacterial restriction enzymes that cleave DNA, often leaving single-stranded ends of specific sequence, known as "sticky ends," that can reconnect to other similarly cleaved fragments. In this interview, Herb Boyer talks about the discovery of the restriction enzyme EcoRI and the early days of recombinant DNA.

Interviewed May 13, 2009

Published September 25, 2009

Once upon a time, not so very long ago, before restriction enzymes were ordered from a New England Biolabs catalog and vectors arrived in neat packages from Promega, and before molecular biologists added patents or a company to their CV, there was Herb Boyer. One can almost define the revolution in molecular genetics by Herb's story alone—the discovery of the iconic restriction enzyme EcoRI and the definition of its sticky ends, the collaboration with Stan Cohen that produced recombinant DNA, and the genesis of the enduring gold standard in biotechnology, Genentech.

I had been fascinated by Herb's story for many years, as I myself had the good fortune to do a post-doc at Genentech in the early 1980s. Much has been written about Herb Boyer, so I chose not to talk with him about his role in the founding of Genentech in 1976, nor the landmark Boyer-Cohen patent, nor the mid-1970s moratorium on recombinant DNA research. Instead, I was interested in what came before all of that—how Herb developed as a scientist, how he become interested in restriction enzymes and in vitro recombination—and by what came later. I'm sure you'll agree, this is equally rich reading.

Gitschier: I know you grew up in the little town of Derry, Pennsylvania. What role do you think your upbringing played in some of the choices you made in your life?

Boyer: My mother graduated from high school, and she immediately married my father at the age of 18. My dad was 12 years older than she. My dad left school after the 8th grade and went to work. He came from a large family with six siblings. He was the oldest boy, so the story was he quit school to help support his family.

I'm trying to write my memoirs now, so I've been thinking a lot about my early childhood. My upbringing was in a town of about 3,000 people and the principal industries were the railroad and a Westinghouse manufacturing facility. I went to a small school, only 32 in my graduating class, and I loved sports and the outdoors. I used to hunt and fish with my father.

Gitschier: Where did your father work?

Boyer: He worked for the Pennsylvania Railroad. My father never owned a car, never had a driver's license. He'd walk to work at the dispatch station, which was about a hundred yards away from our house. He was a brakeman on the freight trains and would live on a caboose for a couple of days at a time. I always thought that was rather romantic. He worked in terrible winter weather and stifling summer heat. By the time he retired, at age 72, he was a conductor, and at that point he was making $12,500 a year. That was my starting salary at UCSF [University of California San Francisco].

High school for me was football, basketball, baseball, girls, hunting, and fishing. And I worked at odd jobs. I dug ditches, mowed lawns for 50 cents, painted houses, distributed door-to-door advertisements, and of course I was a newspaper boy.

Gitschier: Was your father encouraging about your staying in school and going to college?

Boyer: I can never remember getting encouragement from my mother or father to go beyond a high school education, though they weren't opposed to it. I just knew I had to get out of Derry, and that was the only way I knew how to do it!

Gitschier: Your father lived at least 10 years after you became an assistant professor at UCSF.

Boyer: Yes, he was 83 or 84 when he died.

Gitschier: So, he watched this incredible progression in your life. What was that like for you and for him?

Boyer: I'm not sure he had much of an appreciation for what I did.

Gitschier: Did he ask you about it?

Boyer: Not a lot. He was a very quiet man. My dad was just happy I didn't turn out to be a ne'er-do-well. I had a job and I had kept it for a few years and that was good enough for him.

But Jane, how could my parents relate to this? They didn't know what science was or anything about experimental procedures. This is true of many people today, even with substantial educational backgrounds.

Gitschier: Let's shift gears. I'd like to talk to you about restriction and modification.

Boyer: Four or five years ago, Stan [Cohen] and I received the Sir Run Run Shaw Prize in Hong Kong. In my acceptance speech, I recounted some history of restriction and modification and all the little threads that appear to have twisted my career—binary events that seem improbable to have happened. If this event didn't happen, what would have happened? But it happened, so you go to another event. What is that movie with Jimmy Stewart where he's about to throw himself off the bridge?

Gitschier: "It's a Wonderful Life."

Boyer: It's sort of like that—I can give you a few examples. I went to a Benedictine college and I took an elective physiology course taught by Father Joel. We had a brand new, shiny textbook with a blue and white cover. Each of us was assigned a chapter, and we had to give a seminar on it. Which one did I get? "The Structure of DNA." This was 1957, and the buzz of DNA was just getting into the textbooks. And I had this fascination with genetics—classical genetics—*Drosophila*, corn, and bacteriology. I was really taken with the Watson-Crick structure of DNA and this started my fascination with the heuristic value of the structure.

Then I had to decide what to do. I went to the University of Pittsburgh Medical School for an interview with this tough old biochemist. And he said, "Well, if you get into medical school, how are you going to pay for it?" And I looked at him and I said, "You mean I have to pay for it?"

But I didn't get in, which was hard on my ego.

Gitschier: So there is binary point number one.

Boyer: You see! My grades were not that terrific, but remember that was before grade inflation. I got a D in metaphysics, and that didn't help! I was taking a liberal arts program. I got A's in math, logic, etc., but all that other stuff—Chaucer—ouhhh. Someone suggested going to grad school for a couple of years, improving my grades, and reapplying to medical school.

So, I arrived at the University of Pittsburgh at the same time as a professor studying bacterial genetics and gene regulation—Ellis Engelsberg. This is 1958. You know, in 1958 the genetic code wasn't known, and nobody knew anything about messenger RNA, and protein synthesis was still speculative. If you think about what has happened in 50 years—it's unbelievable.

Anyhow, I get singled out because I'm a new graduate student with an interest in bacterial genetics. So I get a chance to work in Engelsberg's lab on the genetic control of the L-arabinose metabolic pathway in *E. coli*. Ellis had recruited Roger Weinberg as an assistant professor and collaborator. Roger had found an L-arabinose mutant defective for the epimerase enzyme. In the presence of L-arabinose, the cell accumulates phosphorylated ribulose, which inhibits growth of the cell. So it's an easy way to select for mutants in all the preceding genes of the pathway.

So Roger Weinberg decides that this should be my project! I had to select mutants induced by mutagens that theoretically could induce specific base pair changes in the DNA. This could be challenged by reversion of the mutation with another mutagen to wild type. After mapping the mutations and then doing amino acid substitution analyses, we would solve the genetic code, the Holy Grail of genetics!

That was the plan! What the hell was I thinking? Why didn't I challenge these guys?

The project required that I map arabinose mutants by the most inefficient way to do recombination that you can ever imagine—P1 phage transduction. At a low frequency, the phage can incorporate small fragments of DNA that when injected into a cell can lead to recombination with the cellular chromosome. Anyway, if you get yields of 10^9 phage per ml you were doing well, and the frequency of recombination is maybe one in a thousand.

So, I started mapping a limited region of an arabinose gene, doing forward and reverse mutational analyses, and I was getting worried, "Am I ever going to get out of here?" But, even before I got too far along on this project, the genetic code is cracked by biochemical means! What a blow! But I continued on with the project for some reason I can't remember.

I did decide on my own, though, that P1 transduction was not the way to do fine structure mapping. In those days it didn't take too much to know the literature, and I knew the literature cold. I was familiar with all the latest work on Hfrs and sexuality in bacteria, so I felt this system [bacterial conjugation] would provide higher recombination frequencies. So I wrote to Ed Adelberg at Yale and asked him if he would send me a couple of Hfr strains. And Ed, being such a super guy, sent them right away.

I started doing the crosses. The strain I had been using was *E. coli* B/r. The Hfr strains, of course, were K12s, so I had to start by asking whether K12 would actually mate with the B/r—no one had ever done it before. I started out by crossing Hfr K12 to the B/r strain and comparing that with K12 to K12, as a standard. I found there was a substantial reduction in frequency of recombination [in the K12 to B/r strain], and the linkage of the various genes was also reduced.

So I started to do some backcrosses and I found out that some progeny of the cross did not exhibit these anomalous genetic results. I mapped the alleles to a region near the arabinose operon. Coincidence!

Gitschier: Did you publish that result?

Boyer: No, not at that time. I managed to write an acceptable PhD dissertation with the other data. However, by that time I was very interested in trying to explain my observation and became more intrigued with plasmids, conjugation, and bacterial sexuality. Ed Adelberg had written a book and papers and reviews on these subjects. So I wrote to Ed and applied for a postdoctoral fellowship with him, and he said, "Yeah, come on!"

Gitschier: Did Ed know about your results about K12 and B/r?

Boyer: Not at the time, but upon arrival in his lab I described my results and he encouraged me to continue my experiments as well as a couple of other projects he suggested. But my heart was in trying to explain my observations.

Ed had a fairly small and close-knit group and we would work in the evenings and chat. Not long before I arrived, Werner Arber and Daisy Dussoix demonstrated that the restriction and modification of DNA, a relatively ignored bacterial phenomonology used for typing clinical bacterial strains, was associated with methylation of DNA and a site-specific endonucleolytic activity. One evening, a graduate student in the lab, Noel Bouck, and I were discussing the paper and she said, "I really think that the anomalies you're seeing are due to the same thing."

So I took out the stains in which I had changed the specificities, and did the lambda phage restriction analyses. And boom, boom, boom, it just lined up. I had mapped the restriction and modification alleles of *E. coli* K12 and B.

I became interested in pursuing it further because of the predicted enzymatic specificity. There were maybe three laboratories in the world working on restriction and modification at that time. I wanted to purify and characterize these enzymes because I thought it would be a great way to study site-specific interactions between proteins and DNA. Unlike repressors, restriction and modification enzymes involved two proteins [the endonuclease and the methylase] that have to recognize the same sequence, and I thought that it would be pretty cool to have two different ways of looking at it.

So, the last year at Yale, I experimented with some way to assay for the K12 and B restriction enzymes. And then I headed to California [UCSF] where we finally settled on an assay based on the sedimentation coefficient of radioactive lambda DNA [sigh!]. This is pre-gels—what a mess! We had a 6-hour run, and we'd do three runs a day. You'd take the little centrifuge tubes, punch them at the bottom, collect the contents drop-by-drop on little squares of filter paper hung on a pin on a piece of styrofoam, dry the papers, and put them into the scintillation counter. It was SO bad. And we couldn't find any activity at all!

Then Matt Meselson and Bob Yuan at Harvard—I can't remember the rationale—they threw S-adenosyl methionine and ATP into the reaction and got activity! So we did that, too, and we started purifying those enzymes.

Gitschier: What year roughly are we?

Boyer: The period of 1966–1969. We went on to purify the B restriction endonuclease and began experiments to determine the sequence at the cleavage site, which we assumed would be the recognition site. So we kept labeling the 5' end of cleaved DNA molecules and we always ended up with equal mixtures of four nucleotides. We never even thought that there would be an endonuclease that would bind at a specific site and then move! I was so disappointed.

Gitschier: But at some point, you make a switch, and you start working on a different restriction enzyme.

Boyer: Well, here comes one of the most bizarre little binary points in life! We found out that these [Type I] endonucleases aren't cutting at a unique site. We used a small phage DNA intermediate and cleaved it with the B endonuclease. By sedimentation analysis it looked like it had a molecular weight of a linear fragment, and that was what threw us off.

Gitschier: Why?

Boyer: Well, we thought it had one site, and you assume it is always cleaving at the same site! Apparently it was cleaving, on average, one site, but not the recognition site. Stu Linn at Berkeley demonstrated, by electron microscopy, that what appeared to be linear products of the B endonuclease could be [denatured and] reannealed as circular molecules.

So I thought, this is it! This is the end. We're not going to be able to determine the [recognition] site with the technology available at the time.

But there was literature, mainly from Japan, demonstrating that bacteria carrying drug resistance factors often had genes for the restriction and modification of DNA with different specificities. So we decided to investigate these enzymes. I had a graduate student who had a degree in clinical microbiology and had experience working in a hospital medical microbiology lab, a wonderful guy named Bob Yoshimori.

I asked Bob to go to the clinical lab at the UCSF hospital and get a slew of multiple drug-resistant *E. coli*. He came back with 36 or so *E. coli* isolates that had multiple drug resistance. And of those, we found eight or ten with restriction and modification activity as determined by phage specificity analyses. We transferred the plasmids into our K12 strains [restriction mutants]. Most of the specificities were like the one that had been reported previously, namely RII [named after the RII plasmid], but we found one that was unique, and that was EcoRI.

We found out later that the EcoRI plasmid came from a woman who was admitted to the hospital with an *E. coli* urinary tract infection that was resistant to multiple antibiotics. Now, the RI endonuclease was never found anywhere else for 15 years, and then it was found in a freshwater microorganism. It just had the same specificity as RI.

Gitschier: Wow, you should write her a thank you note.

Boyer: Well, I wish I had her name. But how would I explain this to her? She's probably not even alive today.

Gitschier: And it could backfire. She could sue!

Boyer: Well, I've thought of that too!

Gitschier: OK, you've got EcoRII and EcoRI.

Boyer: We purified the restriction and modification enzymes of both specificities. We were so thrilled with the first centrifugation experiments. We digested lambda DNA and we had these clear-cut separations of fragments in the sedimentation analysis. And the patterns were different from each other. We went on to determine the sequence of the cleaved and methylated sites of the RI and RII enzymes. This gave us a belated sense of achievement given our prior experiments.

Gitschier: Was it the *Haemophilus influenza* work that was going on around the same time that made you think there might be some enzymes out there for which you could find a cleavage site?

Boyer: We knew about the work of Ham Smith and Dan Nathans. And that was *so* frustrating because there they got that sequence and we had been working for at least a year! By that time Howard Goodman had come to UCSF. He had experience sequencing RNA molecules, and so we naturally started a collaboration on the sequence of the cleaved and methylated sites. At that point, we knew that RI didn't cleave as frequently as all of the other enzymes.

And then Paul Berg called. He had heard about this enzyme and wanted to know if he could get some. I said, "Sure." Bob gave someone from his lab enough enzyme to last a lifetime, and Berg gave it to a couple of his post-docs and graduate students. And it was actually Janet Mertz and Ron Davis, an assistant professor at Stanford, who cleaved SV40 with EcoRI and then looked at it under an electron microscope. They found that the cleaved DNA would circularize at low temperatures, and that's the first evidence for the enzyme generating cohesive ends.

Paul told us this while we were working on the sequence of the cleaved end. We already knew the 5′ nucleotide, so I went to Mike Bishop and said, "Mike, we need some reverse transcriptase," and we filled in the single-stranded part of the end and got the sequence overnight. It was another eureka moment! There was a young medical student [Judy Aldrich] working on a summer research project and she and I looked at the results the next morning, a Saturday—it's GAATTC!

Gitschier: So the RI sticky ends led directly to in vitro recombination. Let's talk about how that took off.

Boyer: My own interest in recombination goes back to graduate school. It was almost an article of faith that DNA would break and exchange strands at any point along the polynucleotide chain. I remember reading Dale Kaiser's papers on the cohesive ends of lambda, and musing about restriction enzymes and Sanger's techniques for determining the sequences of proteins with two-dimensional chromatography. I was thinking if you could break down DNA with these different enzymes, given their specificities, you might be able to separate smaller fragments and sequence them somehow.

I got an invitation to go to Hawaii around 1971 for an East-West conference on plasmids and drug-resistant factors. Stanley Falkow was there—a great guy, known him since my days at Yale. While talking to Stan about our enzyme work, I said, "Stan—you know those plasmids you work with—we can take these things apart and separate the fragments and maybe look at where these resistance genes are." And he says, "You go talk to Stan Cohen—he's interested in that."

So Stan Cohen and I get together and learn of our mutual interests in plasmids and in vitro recombination. He had just described pSC101, which conferred resistance to tetracycline, and we realized it might be of significant value given its small size. And just as importantly, he had become aware of a scientist at the University of Hawaii who could transform *E. coli* with DNA. It made all the difference in the world to our thinking. So we agree that we would see if we could cleave the pSC101 molecule with EcoRI and use it for a collaborative recombination experiment.

Then another stroke of good luck. I was scheduled to go to Cold Spring Harbor to give a talk. I get picked up at the airport by Joe Sambrook and Phil Sharp, and they immediately take me into a darkroom adjacent to their laboratory and show me an agarose gel that had been run with cleaved adenovirus DNA and stained with ethidium bromide. It was one of the most exciting things I could have looked at, and I said, "Thank you, lord!" Because prior to that, we'd have to analyze cleaved DNA fragments by polyacrylamide gel analysis. We would put the gel in a small metal tube and then mechanically push it into a guillotine-like device and slice small fragments into scintillation vials. It would go chop, chop, chop, and invariably pieces would fly across the room and we'd be down on the floor looking for slices. It was like looking for a fallen contact lens. It was so laborious and there was such variation in the tritium counts. So I knew immediately that all the laborious work we had done, we don't have to do anymore!

We immediately found, using the Sambrook/Sharp technique, that pSC101 was cleaved once with EcoRI. Stan sent the DNA up, and we cleaved it and did the ligation.

Gitschier: What was the other entity?

Boyer: Stan had another plasmid with two different antibiotic resistance genes. Annie Chang, Stan's technician who lived in San Francisco, transported the DNA back and forth between our labs. Stan's lab would send us the plasmid DNA, we would do the enzymatic treatments, Stan's lab would do the transformation and selection, Annie would bring back the plasmids, and we would analyze them by cleavage and gel analysis of the fragments.

That was another eureka moment. Bob Helling, a fellow graduate student of mine from University of Pittsburgh who was doing a sabbatical in my lab, and I went to look at the gels in the darkroom, and there it was. It actually brought tears to my eyes, it was so exciting, and I knew what we had done had a lot of potential.

Gitschier: What kind of potential?

Boyer: A lot of my mentors and colleagues were leaving microbial systems to study higher order cells, because "everything there was to know about bacteria was known." But they were frustrated, because they had no hope to isolate single genes or fragments of genes from the chromosomes of "higher" organisms. So when I looked at those gels, I knew we'd be able to isolate any piece of DNA that was cut with EcoRI, regardless of where it came from.

Gitschier: I was just re-reading "Invisible Frontiers" [about the race to clone the human insulin gene, by Stephen S. Hall].

Boyer: Great book.

Gitschier: Yes, and what I didn't appreciate was that October 14, 1980, was the day Gilbert, Sanger, and Berg won the Nobel prize *and* the day Genentech went public. That must have been a very interesting day for you.

Boyer: Yeah. We were gathered at the company to follow the reaction to our IPO [initial public offering] and someone came into the room with the morning [San Francisco] *Chronicle*. And the headline was "Genentech Jolts Wall Street" and underneath is a photo of Paul Berg, "Berg Wins Nobel Prize".

Gitschier: Many people have speculated about why it is you and Stan Cohen have never won a Nobel Prize, but I don't know that you've ever talked about that publicly. Are you going to address this in your memoir?

Boyer: I will, and I don't mind talking about it with you.

It is not for me to decide whether I should or should not win a Nobel Prize. I've received many prizes and honors and I am indeed grateful for the recognition. You can imagine from what I've said about my boyhood, that I never would have expected to do what I've done. I wanted to do something important—I didn't know what it would be, but I planned to work hard and see what I could do. I had no foresight that this would be what it was.

Gitschier: When you say "that this would be what it was", are you referring to the scientific work, or are you referring to Genentech?

Boyer: Both, I don't separate them. I don't know how I could separate those two events in my life.

Disappointed at times? Yeah. But I've been through quite a few periods in my life where I've had strong emotional reactions to one thing or another. All of the criticisms and rebuffs from colleagues that came when I started Genentech, and prior to that attacks on recombinant DNA technology. One of the most difficult periods for me was when my UCSF colleagues were fairly critical; I was the subject of an Academic Senate investigation. Gee, I thought what I was doing was a pretty good thing, and you'd think I was a criminal! That I found to be much more difficult than not getting a Nobel Prize. All in all, these experiences can be of great value to your outlook on life.

I have been rewarded, and I am so lucky, Jane. And I'm so grateful.

You Say You Want a Revolution

An Interview with Pat Brown

This interview focuses on the invention of a DNA "chip," a small surface on which DNA molecules are systematically gridded, enabling thousands of nucleic acid sequences to be queried simultaneously in a single experiment. DNA chips, also known as "microarrays," have been used to illuminate the repertoire of genes active under particular environmental conditions as well as during the development of different organs and tumors. Massively parallel DNA genotyping and sequencing strategies can also trace their roots to this groundbreaking invention.

Interviewed April 30, 2009
Published July 17, 2009

I HAVE KNOWN PAT BROWN FOR ABOUT TWO DECADES and he never ceases to amaze me. Over the years, I have heard him speak quite a few times, and on each occasion I can feel my jaw drop. What will he think up next?

Pat is most frequently associated with the invention of microarrays and their use in studying gene expression, and he should be familiar to the readers of PLoS as a driving force behind open-access journals. But these are only two examples of his many successes, which span the worlds of topoisomerase, HIV integration, protein microarrays, and post-transcriptional regulation. Pat seems to have a brain in overdrive and the energy to match it. I was eager to tap into some of that electricity during the interview.

I met Pat in his office on the fourth floor of the Beckman building at Stanford, where he is a member of the Biochemistry Department. I arrived on a warm and fragrant spring afternoon to find Pat barefoot, wrapping up a grant submission, and obstructed by two large cardboard boxes of assorted PLoS T-shirts. On his door was a small poster: "Where would Jesus publish?"

I knew Pat had an atypical family story, so we started there. He is one of seven talented siblings, who were encouraged by their mother to think big and to make a contribution. His father's work led the family to spend four years in Paris, where Pat attended school in a quaint uniform of shorts with white hat and socks, and a second idyllic stint of four years in Taipei, in a neighborhood surrounded by rice paddies and water buffalo. In between, a Washington, D. C. suburb was home. Pat later discovered that his father did not work for the State or Defense Departments, as he had been led to believe, but rather the CIA, where he was an analyst.

We pick up the interview with discussion of an extremely fertile period in the early 1990s, when Pat was a new faculty member at Stanford and about to launch his work on DNA microarrays.

Gitschier: What was the initial thinking behind the microarray? I understand it had more to do with facilitating genotyping than expression measurements.

Brown: That's right. I was working on a scheme that had the ultimate aim of determining whole-genome genotypes of millions of people for linkage and association studies. It involved a biochemical method that we called "genomic mismatch scanning" for isolating the sequences that were identical between two genomes, and then mapping them by hybridizing to a physically ordered arrangement of the human genome.

At the time, you could map a cloned gene by FISH [fluorescence in situ hybridization] to metaphase chromosomes on slides, and that worked pretty well, but it wasn't scalable for the kind of experiments I was planning to do. You couldn't have Uta Francke, for example, just doing FISH after FISH experiment for all the sequences that we would be generating from this project.

I had a vision about how all this was going to go. I had sent a little blurb to Claire [Weinstock, an administrator at the Howard Hughes Medical Institute] outlining my plan. I used red and green dots to symbolize the microarrays, because I like that color combination.

Gitschier: You obviously aren't color blind. FISH uses just a single fluorescent probe, so why did you feel the need for a two-color, comparative system for the microarrays?

Brown: You need it to make reliable measurements. Kallioniemi had developed a method for complex probe comparative hybridization to metaphase chromosomes for looking at copy number variations. And that was precisely the rationale. If you were to just do a single probe hybridization, you would have very inhomogeneous patterns, only partially driven by the copy number changes themselves, but also by technical factors.

I had a small pilot grant from the NHGRI [National Human Genome Research Institute] to develop the genomic mismatch scanning method, and once Stan Nelson and I had that method worked out, I submitted a renewal application that included the microarrays.

I had a terrible experience with my renewal. In retrospect, I felt it was one of the best grant proposals I have written. And it got the worst priority score of any grant, not only of any grant I've ever written, but any grant I've ever SEEN.

Gitschier: Because it was too ambitious?

Brown: Yeah!

Gitschier: I can just imagine. They probably said each one of these specific aims is an entire grant.

Brown: The specific aims were 1. Take what we've already been doing biochemically [genomic mismatch hybridization] and make it work better and focus on the mammalian genomes instead of yeast. 2. Develop the microarray system from scratch. I said here's how I think we can do it, and it was pretty much exactly as we did start to do it. Aim 3 was development of statistical tools to take advantage of the high-resolution genome-wide genotype. Then I had an aim that we needed to start to put together the infrastructure to do this on a population basis. I had the idea that West Virginia was going to be a good place because it had the smallest fraction of the population moving in and out of the state. And there was yet another aim I can't remember.

I'm just trying to give you a sense of the weirdness of the grant. This is at a stage where all we had really done was to get this biochemical thing working in yeast.

Gitschier: So this was 1992.

Brown: Yeah, we submitted it in November 1992. And I thought, "This grant just totally rocks." And then I got the little note-card back from the NIH [National Institutes of Health]. I saw my priority score: 344. I was just so totally deflated that I literally had to lie down on my office floor for ten minutes to regain my composure.

I got back in touch with them [NHGRI] and they said, "Just do aim 1 and resubmit the grant." I did resubmit, but even in the grant proposal I said, "Following the advice, I'm doing this, but I think it's BAD advice and when I do get the grant I'm just going to be going ahead with other things I had proposed." It was kind of stupid, but I was so pissed off that I just didn't want them to think that I was going to knuckle under.

Gitschier: And did you get that grant?

Brown: Yeah. It was much smaller, but I got it. Meanwhile I recruited Dari [Shalon] to start the microarray work, and that was a strange experience.

Gitschier: Tell me about it.

Brown: I went over to this building called CIS [Center for Integrative Systems], which is where they have a whole bunch of stuff set up, like the n-1 generation from micro-fabrication of the chip industry. So I thought, "This is where all the good stuff is for making very precise tiny things and patterning them." So I went over there and just literally wandered around, asking people who I should talk to and I found Greg Kovacs, who was an MD/PhD neurologist. He was interested in building chips to be used for bionic people, for sensing impulses in nerves and controlling artificial limbs.

I got along well with him and told him what I had in mind to do and how I thought it could be done. I had it all laid out, using robotic printing. He kept wanting to make it a complicated electronic device. But I wanted it to be incredibly simple, and I wanted to use fluorescence read-out, not a circuit detecting changes in capacitance, for example, for a couple of reasons. One—I wanted to be able to do two colors and have the internal control, which you can't do with that direct sensing thing, and two—I didn't want some expensive high-tech thing that was going to be finicky.

That wasn't interesting to him, but he said, "I have a very good rotation student and this guy just wants to work on a project that is practical." What it came down to was that he wanted something that he could use to start a company.

So I thought, "Fine, that's easy!" So I met with the guy [Dari] and told him the main thing I thought it would be good for commercially, which turned out not to be true, was for medical diagnostics. That you could build an array that would monitor the expression patterns in white blood cells. That these cells were acting like spies, that they were circulating to every part of the body—their whole purpose in life was to detect any kind of trouble and orchestrate a response, which involves a transcriptional program. So therefore, you should be able to take a drop of blood and look at what genes are expressed in white blood cells and figure out what they are seeing as an all purpose diagnostic.

Anyway, this is just an example of my attempts to lure him into the project. My pitch wasn't correct, but it had the effect of getting him to work on it. So Dari signs on. And I also had to give him clearance that if he developed something he could then take it and turn it into a commercial product.

Gitschier: Did he physically do this work in your lab?

Brown: Yes, but you had to live with Dari's personality. He was not an adorable guy.

Before Dari came in, I had this whole thing mapped out, an XY robot, we'd have stuff in micro-well plates and just dot spots. I wanted to use a system like a fountain pen because it's simple and robust—500-year-old technology. Dari wasn't too keen on that. He had a lot of ideas of his own; for example, he wanted to print on a linear tape which you'd scan by pulling it through some kind of reader. I never liked this idea at all, but he thought it might be a better system for scanning. There were a whole bunch of ideas, but finally we returned to the capillary printing thing. And that worked fine.

Gitschier: Had capillary printing been done before?

Brown: Not that I know of, but it must have been. The idea is so fundamental.

Gitschier: But you had this idea of a little fountain pen picking up a little bit of liquid, depositing it, going back and picking up something else.

Brown: Yes, the first model I had was from doing electron microscopy. When you pick up a grid, you hold it with these very fine tweezers and you put a little drop of stain on, and this annoying thing happens, not infrequently, that the little drop gets wicked up in the tweezers. And I had been doing some electron microscopy of some virus stuff. So literally the first things we used to do printing were tips of electron microscopy tweezers held together with a little epoxy to serve as our pen. So it was turning this annoying property of electron microscopy tweezers into something useful.

Gitschier: So where did you get the robotics? Did you build it?

Brown: Dari built the first one. Meanwhile, Joe [DeRisi] came to my lab to work on retrovirology stuff. I was trying to get him involved in the microarray stuff because I was trying to shift the center of gravity of the lab, but initially he wasn't buying it. But then, he was in the bay just down from Dari and eventually he got so annoyed that he felt he had to step in and do it better. Joe built the second and third generation printers. His robots are much better and fancier.

Gitschier: Eventually, though, Dari got something gridded.

Brown: Probably in less than a year. I have Dari's thesis somewhere up here. [Shows me.] Here is fluorescent DNA arrayed on the slide, just to show you can do it.

Gitschier: OK, so now we know we have DNA on slides, and we're going to do an experiment. The first one that is published is the Schena paper, which is an expression experiment and *Arabidopsis* at that. It has nothing to do with your original intent of genotyping. Tell me about that turn of events.

Brown: We [Stanford Biochemistry Department] have these yearly retreats. Dari was up to present from my group, and it might have been even in the same session as Mark Schena, who was in Ron Davis's lab. Mark talked about an experiment that he was trying to do. And this was the first talk about the Affy [Affymetrix] array.

Gitschier: So, just a sec. Somewhere in here, Affy is a player?

Brown: They had published a paper on putting optically encoded peptides on chips, a fantastic paper. Then we heard that they were working on doing this with oligonucleotides. And I knew they were at very early stages, able to make only 8-mers. Mark was trying to see if you could use those arrays to look at mRNA expression. But it didn't work at all—you got completely non-specific hybridization.

So immediately after he talked, and Dari had just given his talk, those two guys launched a collaboration, since we had microarrays that were clearly working. We had printed arrays with a bunch of different DNA sequences and different probes. Very high signal to noise.

Mark's idea was to take an isogenic strain overexpressing some transcription factor [and to look at the differences in expression profile compared to control], and I wanted to look at different parts of the plant, but it was all about a cute proof-of-principle experiment more than biology. It was a simple experiment because he had a bunch of cDNA clones and RNA isolated, just a matter of labeling it. Within a month or two, we had data for a paper.

The next interesting paper, as far as I was concerned, was the paper in which Joe was the first author. It was one of my all-time favorite papers. It was what we were going for from the get-go, which is to be able to look at a whole genome.

Gitschier: But this thing with Schena and the expression tipped what kind of questions you were going to ask.

Brown: Right. One thing about the *Arabidopsis* experiment that made a big impression on me was that even by looking at a trivial number of genes, suddenly you could see a picture that is telling you the difference between a leaf and a root.

I just got very excited about it. You don't have to know anything about the mechanism at all. It made me switch gears and made me realize that actually, if you say that genetics is relating variation in the genome to variation in phenotype, there is more accessible variation in expression than there is in sequence, and there is more variation in phenotype between cells and tissues and organs than there is between people. From the standpoint of figuring out biology, that was probably the angle that was going to be more powerful.

What really tipped the balance was Joe's experiment on the diauxic shift, where a whole bunch of things became clear to me for the first time. How powerful it was to look for sets of genes that had correlated expression and how much information that carried about phenotype. Also, the fact that you could take a genome, in which only a third of the genes are annotated, and by looking at their patterns of expression, make pretty strong guesses about what they [the unannotated genes] may be doing. At that point, I thought, I still love genetics, but this is *so* the low-hanging fruit! From the standpoint of doing exploratory experiments and discovering things—it was going to be way more fun.

People in the lab who were doing experiments just looking at gene expression patterns were just turning the crank. For them it was—have an interesting idea for a biological experiment, get data. Genotyping just couldn't compete any more.

Gitschier: I couldn't help but wonder, though, whether at some point the tail started wagging the dog. In other words, have you found yourself in a situation where you were too successful so that your time has been spent, maybe in these bigger collaborations...

Brown: You are so dead on!

Gitschier: ...possibly to the detriment of your own creativity.

Brown: I feel that there is a lot of truth in that because I can get excited about just about anything! In the early days, I thought the best possible thing to do—and I told people in my lab to just roll with me on this one—was to seize any opportunity to get other people to provide us with the best possible samples. Because I thought that a large part of the way we were going to be able to make sense of every experiment we did was to collect a huge body of data. Data would have emergent properties that would make every little bit of it make sense. You could learn the dictionary of how to make sense of how the genome's language was used.

So, I was very promiscuous in terms of soliciting and when solicited, saying yes to collaborations. But what happened was that, very early on, I realized that we had not looked ahead enough. We had tons of things that could turn into papers, but a limited capacity to stop and write papers, especially when a collaborator would take some morsel that was very interesting and want to write a paper on it, but for me it was just one piece of the puzzle. We had experience in turning the data into a story, so we couldn't just hand off the data to people [without our help]. So that became a big drain.

I was a true believer in this and I think it was all very worthwhile. But there is a point at which you sort of know what the answer is going to look like, and where it is headed, and it's very important to see it through, but at that point for me, I'm ready to hand this off.

Gitschier: OK. May I ask about this grant application? I see some tissue staining [on the computer screen].

Brown: It's about developing methodology and software so that you could use a variety of different antibody stains and chemical stains in tissues, and for each one you have a quantitative value, for each pixel you have a vector of values, and you can cluster them the way you do for microarray data, to find things that are similar. And then you color code the images. And in this way, you can actually see specific cell types and pick up subtle quantitative differences in the staining.

One of the things I'm most interested in is that in most tissues, there is a lot more personality to the cells than you think. So I wanted to develop a way to look at a tissue section and say, these aren't just fibroblasts here, but actually 30 different kinds of cells.

This is going to be a really great diagnostic tool too. I want it to be really cheap and fast. You can stain a tissue with a few stains and the code takes almost no time to run. I showed this to Mark Krasnow and he said "We've been trying like hell to find a stain that showed a difference between these two cells," and here we just threw on a few stains with absolutely no specificity for them, but the pixel clustering pulled out subtle quantitative differences in the staining and cleanly separated them. It's just like in FACS sorting [fluorescence-activated cell sorting].

Gitschier: This is very cool!

Brown: Ask me about my next big project.

Gitschier: OK, but first let's spend a minute on the genesis of PLoS.

Brown: I want to LITERALLY overthrow the scientific publishing establishment.

Gitschier: Do you want to say that again, only louder?

Brown: That is what I want to do. PLoS is just part of a longer range plan. The idea is to completely change the way the whole system works for scientific communication.

At the start, I knew nothing about the scientific publishing business. I just decided this would be a fun and important thing to do. Mike Eisen, who was a post-doc in my lab, and I have been brain-storming a strategic plan, and PLoS was a large part of it. When I started working on this, almost everyone said, "You are completely out of your mind. You are obviously a complete idiot about how publishing works, and besides, this is a dilettante thing that you're doing." Which I didn't feel at all.

I know I'm serious about it and I know it's doable and I know it's going to be easy. I could see the thermodynamics were in my favor, because the system is not in its lowest energy state. It's going to be much more economically efficient and serve the customers a lot better being open access. You just need a catalyst to *get* it there. And part of the strategy to get it over the energy barrier is to apply heat—literally, I piss people off all the time.

Gitschier: OK, Pat, with that, I think I'm ready to hear about the *next* big project.

Brown: OK—I'm serious, and I'm going to do my sabbatical on this: I am going to devote myself, for a year, to trying to the maximum extent possible to eliminate animal farming on the planet Earth.

Gitschier: [Pause. Sensation of jaw dropping.]

Brown: And you are thinking I'm out of my mind.

Gitschier: [Continued silence.]

Brown: I feel like I can go a long way toward doing it, and I love the project because it is purely strategy. And it involves learning about economics, agriculture, world trade, behavioral psychology, and even an interesting component of it is creative food science.

Animal farming is by far the most environmentally destructive identified practice on the planet. Do you believe that? More greenhouse production than all transportation combined. It is also the major single source of water pollution on the planet. It is incredibly destructive. The major reason reefs are dying off and dead zones exist in the ocean—from nutrient run-off. Overwhelmingly it is the largest driving force of deforestation. And the leading cause of biodiversity loss.

And if you think I'm bullshitting, the Food and Agricultural Organization of the UN, whose job is to promote agricultural development, published a study, not knowing what they were getting into, looking at the environmental impact of animal farming, and it is a beautiful study! And the bottom line is that it is the most destructive and fastest growing environmental problem.

Gitschier: So what is your plan?

Brown: The gist of my strategy is to rigorously calculate the costs of repairing and mitigating all the environmental damage and make the case that if we don't pay as we go for this, we are just dumping this huge burden on our children. Paying these costs will drive up the price of a Big Mac and consumption will go down a lot. The other thing is to come up with yummy, nutritious, affordable mass-marketable alternatives, so that people who are totally addicted to animal foods will find alternatives that are inherently attractive to eat, so much so that McDonald's will market them, too. I want to recruit the world's most creative chefs—here's a *real* creative challenge!

I've talked with a lot of smart people who are very keen on it actually. They say, "You have no chance of success, but I really hope you're successful." That's just the kind of project I love.

Do you feel like you are ridiculously optimistic?

Gitschier: Me? Yeah, sometimes. I have my share of wild ideas. But you—you want a revolution.

All About Mitochondrial Eve

An Interview with Rebecca Cann

The mitochondrion, a cellular organelle that generates energy by oxidative metabolism, contains its own small circular genome. During the reproduction of animals, all mitochondria are contributed by the mother's eggs. Consequently, mitochondrial DNA is inherited exclusively through the maternal lineage, and the mutations that accumulate in mitochondrial DNA from generation to generation can provide a genealogical record of maternal ancestry. In this interview, we learn how tracking mitochondrial DNA, purified from people with ancestral roots throughout the world, first demonstrated that the lineage for all modern humans has its roots in Africa.

Interviewed October 20, 2009

Published May 27, 2010

IN UNEARTHING THE GENETIC HISTORY OF HUMAN POPULATIONS, the recent pace of discovery has been so rapid that we can lose sight of the impact made by a single paper. In a 1987 *Nature* article, Rebecca Cann and her co-workers, Mark Stoneking and the late Allan Wilson, painstakingly analyzed mitochondrial DNA purified from placentas that had been collected from women of many different ancestral origins. By comparing the mitochondrial DNA variants to each other, the authors produced a phylogenetic tree that showed how human mitochondria are all related to each other and, by implication, how all living females, through whom mitochondria are transmitted, are descended from a single maternal ancestor. Not only that, they localized the root of the tree in Africa. The report left a wake, still rippling today, that stimulated not just geneticists and paleoanthropologists, but the layperson as well, especially as the ancestor was quickly dubbed "Mitochondrial Eve." Indeed, the cover of *Newsweek* one year later depicted an Eden, replete with apple tree and serpent, but with the iconic blonde couple of Dürer now supplanted by an Adam and Eve of African descent.

I have always marveled at this paper, particularly as I had gotten to know Becky Cann when she was in the throes of completing this study and writing up the 40-somethingth draft for its publication. At the time, I didn't appreciate the magnitude of what Becky had accomplished or the implications of the work. When the American Society of Human Genetics meeting was held in Honolulu in October, I arranged to interview her in her lab at the University of Hawaii to elicit her reflections on this discovery.

We met in the late afternoon, as students came by to submit their term papers under the 4 p.m. deadline. Becky has a kind of earth-mother quality about her, plain-spoken, clear-thinking, and supportive of her students. Still active in the field of human origins, Becky has also branched into avian work, examining the phylogeny of endangered species of native Hawaiian birds in Hakalau Forest National Wildlife Refuge with her husband and collaborator, Lenny Freed. The University's concrete campus lies deep within a canyon, its hard edges softened by mist and vegetation. Becky seems to fit right into the environment, fully grounded and far enough from the fray to follow her own path.

Gitschier: I noticed on your CV that you went to high school in San Francisco. Did you grow up there?

Cann: No, my parents moved from Des Moines, Iowa to San Francisco the summer before I started high school. My parents thought that they would gently move us into California by putting us in an all-girl Catholic High School, and they sent us to one on the edge of the Haight-Ashbury, not realizing the neighborhood it was in at the time and what we would be seeing on the way walking to school. It was quite a trip!

Gitschier: I'll bet—this was in the late '60s, right?

Cann: Yes—1967. I had a number of people ask me if I would like to try their pharmaceuticals on the way to school. They thought it was really funny to try to convince a little Catholic high school girl in a blazer, blue and white saddle shoes, a plaid skirt, and a white blouse to sample the wares.

Gitschier: But you made it through the school relatively unscathed—

Cann: Well, I think there are very few who make it through Catholic school, the last four years with only girls in their teenage years, unscathed.

Gitschier: OK, so it wasn't the Haight experience that was the problem.

Cann: I have a nun that spits fire on my windowsill.

Gitschier: Still! I had assumed from your last name that you might be Jewish.

Cann: My ex-husband was named Cann. I married for the first time in 1972, right after I graduated from Berkeley with my Bachelor's degree. I put an ex-husband through graduate school and then when he finished, I started graduate school.

Gitschier: Yes, I saw that you worked at Cutter Labs for five years after college. What was the evolution of that?

Cann: I got a Bachelor's degree in Genetics, and I thought I was interested in the genetics of behavior. I wanted to work on primates or humans, thinking that it would be a lot more interesting than fruit flies. But by the time I graduated, I came to see that the tools for doing human genetics were pretty crude. This was 1972.

Gitschier: Well, there was...

Cann: Nothing! Even restriction enzymes weren't there yet! So, I decided that since I really didn't have a good idea of what I wanted to do, I would find a job working as a quality control chemist—because I had the course work and I could get hired. Initially I began working nights so that I could take classes at Berkeley and read during the day. I used that time to branch out and think about how to put that interest in humans and human populations into a genetic context and to work explicitly at a molecular level.

I worked at night, right next to the mailroom, and I'd see all these journals coming in. Company scientists were talking about individuals varying in their ability to metabolize given compounds and how this influenced dose-response curves and that drugs that were developed for one population might not be effective for another. This woke me up to the fact that there really was this thing of personalized genomes that have a phenotypic effect. But how to get at the genotype?

I began to realize that there really was a significance for understanding those kinds of questions. But I was just learning how to be a good lab biologist at that time—how to be accurate, keep records, learn new technology, reproduce findings, and not to be frightened by the repetition of science.

Gitschier: That stood you well, then.

Cann: Oh, god—147 placental DNA preps later—yes!

Gitschier: OK, something makes you decide to go back to graduate school at Berkeley....

Cann: Yeah—restriction enzymes! I'm reading about them and taking classes, I learn about this crazy guy—Allan Wilson. Allan has this crazy idea that you can measure evolution by measuring mutation, but, in order to do that, you have to be able to find the same piece of DNA reproducibly in different individuals and in different species, so that you can do that kind of "molecular clock" comparison.

And he had already collaborated with an anthropology professor, Vince Sarich. They had tried to do this indirectly with immunology, to look at the degree to which you could force an immune response. The idea being that the more closely related two individuals are to each other, the more parts of the immune system they would share. And with this technology they produced these trees that caused the paleontologists to just go nuts!

Then, suddenly there started to be this inkling that if you had a restriction enzyme you could do a Southern blot, and if you could do a Southern blot you could now start doing more individuals. You would be looking only at those genes for which you had a good probe, but you could do it a lot faster than you could produce individual amino acid sequences and there would be more variation there.

So, at that point, I decided this was going to be the wave of the future. There was a change in biotechnology, and I was goin' to grad school!

Gitschier: And you knew you wanted to work with Allan.

Cann: Allan or Vince. I didn't realize the degree to which they shared the laboratory at that point. And Vince was a real bench scientist—Allan wasn't. He had left the bench long ago. We used to joke that he didn't know which end of the Pipetman to put in the liquid. Allan was a professor of biochemistry and Vince was a professor of anthropology.

Gitschier: So, were you a graduate student in biochemistry?

Cann: This became an issue. I had been admitted to anthropology, and there was this question of whether I could function at the same level as biochemistry students. So there was this elitism right away. But Allan, once he watched me at the bench and knew something about my history, dismissed it. [I was] somebody who had worked at the bench as an industrial chemist and whose livelihood depended on doing the right thing. Very soon I was in the position of teaching his younger graduate students how to do bench science, and it was a big lab. At one point, Allan had 18 graduate students and six to eight post-docs. There was a lot of competition both for his time and also bench space. You had to be productive, 'cause otherwise Allan wouldn't talk to you.

Gitschier: What year is this?

Cann: 1977. I was working on macaque serum proteins doing PAGE electrophoresis and stacked gel electrophoresis, looking at macaque species systematics—the shape of their phylogenetic tree. These are old-world primates that have spread out throughout the tropics and as they became isolated, they speciated.

Gitschier: So you're running serum proteins on a gel...

Cann: And you're staining with Coomassie blue, and you're asking what bands are shared between species. But you don't know which proteins are which.

So then we thought, let's use some hemoglobin Southern blots and let's see if we can sort this out. The idea of doing something very specific [with DNA blots] was very appealing: looking at the degree to which restriction maps matched or didn't match.

And then Wes Brown blew in. He was a post-doc from UCSF, with a PhD from Cal Tech in Jerry Vinograd's lab. They did the original mitochondrial DNA isolations and had been isolating small viruses to chemical purity using CsCl density gradient centrifugation. Because Vinograd had just died, almost before he finished his thesis, Wes moved up here and was coming over to write grants with Allan. He knew that Allan was interested in timing and clocks and evolution. He said, "By the way, you know you could take this purified fraction of mitochondrial DNA away from the genomic DNA and then bust it open with restriction enzymes and study that—and this stuff changes really fast!"

So Wes and Allan were writing these grants, and Wes had a paper on 21 isolated mitochondrial fractions that he had gotten from placentas from labor and delivery. And they showed this incredible variation! But the medical records were disassociated; he didn't know much about the donors [i.e., their ethnicities].

It got to the point where you could end-label [DNA fragments with ^{32}P]. So he was taking Klenow fragments, end-labeling his pure fractions after restriction enzyme digestion, and running them out on long gels to construct a physical map. You'd figure out the size of the fragments, knowing the total size of the [mitochondrial] genome, starting with the 6-base cutters, and map the mitochondrial genome.

Wes's big contribution was totally changing over Allan's lab from a protein-biochemistry approach to studying evolution to a DNA-based approach. He brought not only the restriction technology, but also cloning mitochondrial fragments. We did Sanger sequencing and showed exactly the spectrum of mutations, and that even though there were essential genes in the mitochondria, they still changed faster than the same classes of genes in the nuclear genome.

Gitschier: What prompted *you* to make the switch to mitochondria?

Cann: I saw that that was a potential tool to break open the question of human variation.

All this time I was taking graduate seminars. I took all the human anatomy, taught by the physical anthropologists and Tim White, who was part of the Laetoli footprint team in the late '70s. He was the anatomical expert during the Lucy discovery, too.

So I'd come from the lab, and I hear all this "yammer yammer yammer" about 2- and 3-million-year-old fossils and which lineage goes where. And from my understanding of the human fossil record, we hit Cro-magnon and nobody knows how that's related to Neanderthal. And *Homo erectus*. What happened there?

So talking to Allan and to Wes, and reading and arguing a lot, [we thought] potentially you could expand this view of human evolution in that time period by doing a mitochondrial analysis, because there was enough variation there. If you used your average nuclear gene, there wasn't enough resolution. The thinking was that this technology would give us enough differences between populations to start asking whether there is a most-recent common ancestor that is different for this group vs. that group, and how these older archaic populations in Asia and Africa are related to the modern people.

Gitschier: But did you imagine that you could actually get mitochondrial DNA from the archaic people?

Cann: Why not? Some of the fossils I was looking at had organic material in them. And people had been trying to get stuff out of blood on tools. So, who knew what you could get out of a fossil? At the same time, Allan came back from sabbatical and he had a little piece of a frozen Siberian mammoth that they had on exhibit;

he had shaved off the heel. And he had people going to the gem shows to pick up amber.

Once the mitochondrial thing took off, the reason people started thinking about ancient material was that we knew mitochondrial DNA was such a large fraction of DNA in the cell, so we knew if anything was going to survive, it wasn't going to be single copy nuclear genes—it would be something in the mitochondria.

It was a fertile time. Lots of ideas floating. Allan used to say, "Keep having ideas—some will be good, some will be crap. Just keep thinking them up."

Gitschier: OK. I want to talk to you about a paper by Cann and Wilson in *Genetics* in 1983. You've made DNA from 110 human cell lines or placentas, including from people of African descent, and you have this tree. But this tree doesn't look like the next tree in the big paper [*Nature* 1987]. You don't show African origins in the 1983 paper; in fact, you say there is no strong correlation with race or geography, consistent with multiple origins for length mutations. This ultimately is not correct, is it?

Cann: Unh unh....

Gitschier: So, what I'm trying to figure out is what happened between these two papers.

Cann: PAUP, the new phylogeny program—"phylogenetic analysis using parsimony"—devised by David Swofford. In the earlier papers, we used Fitch-Margoliash trees, which are *distance* trees.

There are two ways that you can draw a pattern of relatedness between two individuals. One way is to look at similarity, and just take that measure—just add up all the differences.

Gitschier: So let's see if I get this. In the 1983 paper, you have a large matrix of differences between the 110 people...

Cann: [Nods] Or, instead, if you have DNA sequence information or a good enough restriction map, you can see what base had to have changed. The other thing that had happened was the Cambridge reference sequence [for mitochondrial DNA] had been published, so I could take my restriction map and match where the restriction enzyme site was on the reference DNA [sequence]. I could figure out—in order to generate this site or lose this site—what that change had to be. Suddenly, you went from being able to extract not just how different or similar two individuals were and put them on a distance tree, but also, with the sequence, you can use a parsimony principle and say what changes are present in two or more individuals.

You are trying to use the information with the assumption that the mutation happened once and only once. And then successively build up these blocks of sequence that have to be more closely related to each other, with that assumption, by parsimony.

Originally the PAUP algorithm couldn't take so many samples—originally it was 30 by 30, and then 50 by 50. And by the 1987 paper, PAUP could take 150.

Gitschier: And you have 147 samples.

Cann: Yes, in my thesis, in 1982, I had used PAUP along with Fitch-Margoliash and Neighbor Joining—the distance type of algorithms—and tried to compare them. I had to use random sub-draws of 30 individuals with the parsimony tree—because that was the biggest matrix PAUP would take—and I would try to see whether randomly generating those matrices with parsimony gave me something different than the distance trees.

And they did, but I couldn't prove it, just randomly pulling and having variation so that the ethnicities were stratified across 28 Asians and 2 Africans, for example. I was still getting a pattern that was different from what the distance trees gave. But I couldn't convince Allan that it was really showing the African origin. I couldn't convince him.

[Luca] Cavalli-Sforza had published a paper with Doug Wallace saying that Asia is the origin. Other people, like [Milford] Wolpoff in Michigan, were arguing that the mitochondrial trees couldn't possibly ever be right, and I, in particular, could never be right 'cause the greatest human geneticist living had just published this other tree that showed human ancestry in Asia.

Gitschier: Where did they go wrong with that?

Cann: They were using distance trees. They had a really long branch to the Africans in their sample, but Cavalli believed that human origins were Asian, and that Africans just had this wild mutation rate because their environment was so bizarre. [He thought] if you are really going to root the human tree, Asia was a better place to do it.

There really wasn't good evidence, other than thinking that modern humans couldn't have evolved in Africa. Biologically, they had evidence that should have placed it in Africa. And the Japanese geneticists, like Masatoshi Nei, saw that and called it. They said, "You can't defend this. There is no good archeological evidence that would force you to put that root there [in Asia]."

Gitschier: So these samples that you used were the same ones that you had looked at before.

Cann: There were some additions. I got additional Australian aborigines, finally. I had been waiting for about 40 more to come. It was now 1984, I was in my first post-doc and I was cloning and sequencing some of them. Mark Stoneking came to the lab and Allan suggested that Mark do this mapping on the additional Australians and the Papua New Guinea highlands, which was his thesis. So, Mark contributed those to this paper.

Gitschier: At what point is Allan on board?

Cann: We wrote the paper and submitted it in late '85, and it got held up in review for over a year in *Nature*, because the Brits didn't want it to be published.

Gitschier: Why not?

Cann: There was a certain group that wanted to publish their phylogeny of globins.

Gitschier: Ah. Why didn't you just pull it and sent it to *Science*?

Cann: I think Allan wanted the prestige of *Nature*. He's a New Zealander, proud of it, and wanted to show them up. He published a lot of papers there. We talked about moving it, but he had faith that eventually they would understand the value of it. I think he was reticent to talk about the personalities of the people involved.

Whenever I'd get a review back he'd say, "Don't worry about who reviewed it. It's not a positive thing to be thinking about, and it won't help you make a better paper." He knew there were some really sharp personalities that were directed at a woman—an American—and an upstart Colonial. He didn't want me to start thinking that this is what I was going to face in science.

He was a real Marxist. He knew how hard it was for women to get going in science. He'd seen it—his lab was a haven for the women in the program and the male professors from other labs would joke that Allan had all the women. What was so special about Allan? He was gender blind. If you had a good idea, it was a good idea.

Gitschier: But back to the tree—you already suspected that the tree was going to look this way.

Cann: I didn't know for sure that this was how it was going to look. Mathematically, given all these samples, there were lots of possibilities. A universe of trees! But this tree could be reproduced—the order of entry could influence the outcome, so we would reorder the entry, and run against subsets of individuals. This was a plausible tree.

Eventually Allan was comfortable with the idea and that it could be defended. We didn't go out on the limb and say it was the *best* tree, but it was a tree with a high likelihood of being correct and it was consistent with a lot of other data—anatomically modern fossil forms in South Africa around

200,000 years ago. Then White found another fossil in East Africa from around the same time. And the Middle East fossils were re-dated at 110,000 years.

Gitschier: When it came out, this 1987 paper must have changed your life.

Cann: Not for the good, sometimes. I got a lot of hate mail, crank mail, some with strange scrawling notes. I even got a visit from the FBI after the Unabomber attacks. I got random calls in the middle of the night, and people on flight layovers wanted to talk. I was unprepared for this role as the molecular person questioning the fossils—and for people like Wolpoff saying these archaic people evolved into modern people, or that I had studied African Americans, not real Africans...

It made me mad because people were doing the same thing with birds and lizards and fish and they weren't taking anywhere near the amount of crap I was taking. I could see it was only because I was talking about humans. These arguments raised so much emotion, and that really depressed me.

Gitschier: What was Allan's reaction to the press on this?

Cann: He was bemused. People had two reactions: either (1) they knew it all along, or (2) it can't possibly be right. So he was trying to find a predictor of who was going to say what to him. Would it correlate with any other prejudice he had based on past interactions or personality type?

I remember a discussion over dinner one night about three years after it was published. There were a number of population geneticists, like Alan Templeton, who still haven't resolved in their minds that this African origin idea could be correct. He continues to write stuff about this. And it's not that I don't want to listen to criticism, because there were things that obviously we didn't have the answers to when this was written. Mitochondrial sequencing has shown certain areas that will generate distortions, and we didn't have all the samples we would like to have had. There were some leaps of faith.

Sometimes I've heard Luca talk and say, "Well they got the right answer, but they didn't know why they got it." And I always thought that was dismissive. I had a pretty good idea why I got the right answer!

Curling Up with a Story

An Interview with Sean Carroll

Molecular tools have enabled biologists to gain purchase on the age-old problem of how an animal develops from a single fertilized egg. The fruit fly Drosophila melanogaster *is particularly suitable for dissecting questions of body-pattern formation, in part because many mutants that alter development have been described over the past century. This interview discusses the success of one such tool in the fly embryo and the value of comparable studies in other species to understand how subtle variations in DNA govern the evolution of body patterning.*

Interviewed April 11, 2008

Published October 31, 2008

To meet SEAN CARROLL ON HIS HOME TURF in the early spring of Wisconsin is like encountering a bear cuddled up in his lair, waiting out the cold winter. I burrowed into the softly lit cave of small offices, with stalactites of yellow post-its dripping from every imaginable surface. Tiptoeing over misaligned stacks of books and reprints, I had to resist the urge to pick up one of the worn works, settle into a corner, and join in the reverie.

Carroll is an expert in the field known as "evo devo," an amalgam of developmental molecular biology as applied to the workings of animal evolution. Following his initial work with *fushi tarazu (ftz)*—one of the segmentation genes in the *Antennapedia* complex of *Drosophila*—he has been instrumental in elaborating the developmental regulation and interaction of a variety of genes, at first in the developing embryo, and later in the genesis of leg and wing appendages. A chance encounter fueled his long-standing interest in evolution and prompted him to re-tool his lab for the study of butterfly wing development; comparison between the two species led to groundbreaking insights into the subtle evolutionary changes that can give rise to spectacularly different appearances.

Carroll now leads a double life, and what captured my attention was his newfound voice as a writer about evolution, with three books already in print and, as I learned during the interview, two more ready for publication in 2009. We got the ball rolling by recalling how we had been introduced in Boulder, Colorado, while he was still a post-doc with Matt Scott, and I began by asking him about that period of time.

Gitschier: What took you to Matt's lab?

Carroll: Reading as a graduate student. I was an immunology graduate student at Tufts Medical School. I was even thinking that the evolution of the immune system was something to work on in the long term. But in those days, it took a lot longer to run gels, and you had time to read! So I read a lot, and I made use of the Red and Green Lines, getting around to all the schools in Boston. I went to seminars routinely at Harvard Cambridge, Harvard Med, MIT, and Tufts. And I went far afield, often, if it interested me.

Gitschier: What kinds of things did you read?

Carroll: All sorts of things—general science, general biology. Books by Stephen Jay Gould or his *Natural History* columns. History of science. Intense periods of science—atomic physics or cracking the genetic code. I had a strong appetite for that.

I had a growing awareness of issues and questions in evolution. At the time [early 1980s], punctuated equilibrium was a topic being discussed around Boston. And I thought a lot of this debate was about the evolution of form, about how quickly things could happen, and about the genetics of that. I realized you really couldn't have that debate without knowledge of what the genetics of form really were and without understanding how things were really built.

And that persuaded me that the next big step in evolutionary science in that vein was going to require an understanding of the genetics of animal development.

Gitschier: That was so specific!

Carroll: It was a distillation of a lot of cross-currents.

I looked around at what was going on. I came across two papers—one was the classic Ed Lewis review in 1978 on homeotic genes and the second was in 1980 by Nüsslein-Volhard and Wieschaus, which is the report of the big screen in flies for zygotic mutants.

There were some whispers that things were starting to be understood molecularly, and that led me to the small group of labs that were working on fly developmental genes. One of those new labs was Welcome Bender's at Harvard. He said that he wasn't taking any more people, but he told me about Matt, who was wrapping up his post-doc work in Indiana with Thom Kaufman. I had some familiarity with Boulder, Colorado, and I thought: couldn't be the worst thing in the world to do post-doc in Boulder and work on these genes!

The work Matt had done as a post-doc essentially set the buffet. He walked through the whole *Antennapedia* complex but had not had time to work on any individual genes—how they were encoded, expressed, regulated.

So when I got to Boulder, it was open season on these genes.

Gitschier: Were you the first person in Matt's lab?

Carroll: Allen Laughon and I were there for day one in Boulder. Allen came from Ray Gesteland's [lab] in Utah. We took over a lab from a microbiologist, and Boulder hadn't bothered to clean it. So Al and I spent the first few days emptying reagents from old bottles and re-filling them with new ones.

We had a DNA map of the *Antennapedia* complex. The whole region, a few hundred kb, was cloned. Breakpoints of *scr* [sex-combs reduced] and *ftz* mutants were mapped.

I had an immunochemistry background, so I had a lot of experience in producing, purifying, and using antibodies. So I had something to bring to the table, but I had never worked on flies. The idea was to localize these gene products during development.

Gitschier: In your first book, you talk about this frustration of 1.5 years of work, and then coming out of the darkroom—

Carroll: Today is the anniversary—April 11, 1985—I even know the day!

Gitschier: I'm so honored to be here! So, what was the experiment? You were trying to localize *ftz* protein in the developing fly embryo.

Carroll: Well, it was really hard to know the path to take. In vitro, you could characterize an antibody and know that it was reacting with an antigen. But the methods for localizing antigens in embryos and imaginal discs were still emerging. A couple other labs were having some success. There was antibody to *Ubx* [*Ultrabithorax*] by that time. Tim Karr was working on fly embryos and had some protocols.

There was a lot of groping—a lot of lore about what vectors to use, β-*gal* fusion products, producing enough antigen, stability problems, purifying the antibody, how to permeabilize the embryo.

You didn't know if there was going to be enough antigen to see! I remember that was a criticism with Matt's grant: how do you even know there is enough protein to detect?

Gitschier: Well, you don't know!

Carroll: You don't know, and that's why we call it *"research."*

The ultimate test was incubating the embryos with antibody and fluorescent secondary antibody and seeing! I don't know how many times that experiment failed in my hands. I devised a different way of purifying the antibodies in larger quantities, in bigger batches, in cleaning them up. I remember thinking, "I can't think of any better way to do this!" I was a year and a half into this, and I wasn't sure that I had any more tricks up my sleeve.

But then—it worked! It was early evening, hitting the scope, and just seeing green stripes [fluorescein-conjugated secondary antibody reacting with the primary antibody revealing *ftz* antigen in seven nuclear stripes]. In whole mount, it was a gorgeous thing to see!

Up to the time, people were doing in situ hybridization to sections and then exposing to film, and you'd have to wait for these things to develop for days and days—then you'd see the silver grain [deposits]. Ernst Hafen in Walter Gehring's lab had caught a nice tangential section that gave them a lot of the stripes. So, stripes of *ftz* RNA had been seen.

But there was something beautiful about seeing the nuclear protein in seven stripes. And I was looking at a pot of embryos that were *all* striped.

Gitschier: That must have been thrilling!

Carroll: Matt was home for dinner, as I recall. He came back in. And there *was* drinking. OK, I was drinking; Matt wasn't drinking.

Gitschier: It's too bad the published article itself doesn't show the color.

Carroll: No, in those days the articles weren't in color. But we did have the cover in color [together with work from Steve DiNardo and Pat O'Farrell on *engrailed*]. It was really brutal to get color images, for color slides and color prints—the cameras were mounted on the scope—they weren't digital, so you'd have to leave the shutters open for 30 seconds to get these pictures—and of course you're bleaching the embryos as you did that. The black and white images you could develop yourself in the lab, but the color stuff you had to send out and wait days to get back.

What the *ftz* and *engrailed* antibodies allowed us to do was to work out regulatory hierarchies. You had a batch of 20 or so loci that affected a segmental pattern—the gap genes, the pair-rule genes, and the segmentation genes. You had all these phenotypes, but you didn't know who regulated whom. The antibodies gave us tools to work this out pretty quickly. Rather than waiting for silver grains and the fortunate section, you'd stain a pot of embryos from a cross of a zygotic mutant line and you've got hundreds of mutant embryos—you've got a clear picture of whether gene expression is or is not altered. And bang! These reagents just sped up the analysis of regulation in space. And the resolution was great—cell-by-cell patterns of gene regulation, tips you got from spatial relationships of expression. Resolution in fluorescence microscopy is superb.

I remember people saying this could never be worked out—you had all these genes working very closely in time and in spatial patterns. You had to work on little pieces of the network. And that got into how individual genes were regulated.

Gitschier: As a post-doc, were you able to read as much as you had as a grad student?

Carroll: No, it was a lot of writing—pipette in one hand, pen in the other. It's going to sound awkward, but from the moment we saw stripes, there was a lot of writing! And writing takes time.

Gitschier: Obviously now you are a very prolific writer.

Carroll: Yeah, I've been de-repressed.

Gitschier: So was this instinct to write under some kind of repression that you weren't aware of? Did you *know* you liked to write? Had you been writing poetry or fiction, or keeping a journal?

Carroll: No! The only thing was that I took a second major, in French Literature. When I went to Washington University as an undergrad, I had to take a French class. And I thought "One French class, I'll bear it," but the professor was fantastic. And I took five more classes with him after that including a graduate course, reading Rousseau, etc. You had to write for that—15-page-long term papers! To write in a foreign language and to write analysis of literature—somehow that was calisthenics for the writing brain and the writing voice.

But writing for science journals—there is a certain amount of DRYNESS to it that is ENFORCED by RIDICULOUS pressure. Did I say that loud enough? But writing of scientific papers requires a lot of discipline, a lot of logic, organization, succinctness.

Gitschier: Who was this inspirational French teacher?

Carroll: James F. Jones, but he goes by Jimmy Jones. He is now president of Trinity College in Hartford, Connecticut.

Last year, I was giving a public lecture at the American Museum of Natural History in New York. And he hired a bus and he brought faculty and students to the lecture and took me out to dinner after the lecture on Broadway. Thirty years later!

That's the caliber of the people at Wash U, and what amazes me now, from the position I now sit in, was how they made themselves available to the undergrads. I now realize that I was a pain in the [neck] and they never said so!

Gitschier: OK, you were probably going to use the word "wing development" before we digressed.

Carroll: Yes—back to the appendages. So my brain was saying "OK, we're movin' out, we're still thinking evolution"—but by late the 1980s, I still haven't done anything explicit yet about evolutionary biology. I'm still preparing with developmental biology—understanding how to make a fly before we start thinking about other things and how they're different.

My early fire was *diversity*! So I wanted to study other animal models that would allow us to exploit what we had learned in flies and pursue questions of how diversity arose.

Then, a critical thing happened. I visited Duke University and I met Fred Nijhout. Fred was interested in endocrinology and a lot of other things—he had discovered the organizing center for the eye-spots in the butterfly wing by classic transplantation experiments in the imaginal disc.

I was talking about bristle patterns on the adult fruit fly, and Fred said "Do you think any of these genes you're studying could draw these kinds of patterns [on the butterfly wing]?" And that was the right question. And I said, "Yeah I think they could, so let's go find out."

I decided butterflies were the right model to start asking questions about divergence and diversity. Butterflies have large hind wings, whereas fruit flies' second set of wings is the haltere. The scales on the wings were different—they are modified bristles. Their geometric color patterns were something new. And butterfly caterpillars have pro-legs on their abdomens. So all these are differences with respect to the body plan. And we probed all those differences.

That was the switch into the evo part of the evo devo for me—and that kind of flew out of control!

Gitschier: Obviously fruit flies are a lab animal, but butterflies? How did you gear up for that?

Carroll: Fred had a colony going for a long time. He sent us the butterflies, the recipe for the food. We learned from Fred how to grow them, so we had a constant supply of eggs in all developmental stages.

We made cDNA libraries, developed tools for in situs of embryos, made antibodies. Wing discs of butterflies are a lot bigger than fruit flies', so this was tricky getting them to look really nice when we probed them.

We cloned all the homeotic genes, the wing-patterning genes, and that gave us our early results. We posed very simply binary questions, and we got answers that were visual and that anyone could understand when they saw them.

It was about 2 years of technical investment before we started to get cool results. For example, of all the genes we study, one was used in a novel way—*distaless*, in the development of the eye-spot. Because it was this ancient gene, used in building legs, and it had taken on this new role, it was a striking, and at the time, I think, the first evidence of any kind of using old genes to make new patterns.

And the other thing, which I wasn't prepared for, was—goodness! How people like butterflies! Some of the public press things started because of butterflies.

Gitschier: When I first read about the butterfly work, I thought, "This is probably a guy who captured and pinned down butterflies as a child." But then I read somewhere that you were into snakes!

Carroll: Yes—but it was all about color patterns!

Gitschier: Well, the butterfly stuff was really pivotal for you.

Carroll: It drew talent to the lab.

Gitschier: And it gave you some opportunities to try your hand at writing some news and views.

Carroll: When either a lot of data are emerging or it is a confusing situation and there is a need to distill, I like that challenge. In 1990, just as there was a sense of how periodic patterns were made in the embryo, I wrote a review for *Cell* about stripes. It was coming out, from Mike Levine's work, that inter-stripes were being repressed and stripes patterns were being carved from a block of potential expression by repressing expression in the inter-stripes. That article was the first effort on my own to try to get somewhere new conceptually.

Gitschier: How long did that take you to write?

Carroll: Months—anything takes me months. I can't even write a postcard in under a week. The re-writing, the honing, trying to draw figures that are helpful.

Then I started doing that more often, especially with the evolutionary stuff. In 1994, I wrote something for a meeting contribution—the first modern evo devo meeting, in Edinburgh.

Gitschier: Who coined the term "evo devo"?

Carroll: Don't know. I don't actually even like the word.

Gitschier: But it's the title of your book!

Carroll: It's the *subtitle* [of *Endless Forms Most Beautiful*]—my publishers like "evo devo". I'm ok with it now.

By 1995, there were some misconceptions about homeotic genes and there were some new data, so I wrote a review for *Nature* in 1995. For a *Nature* audience, you've got to be aiming for those general themes, themes that have a root in history, what people had said before and how data were weighing in on long-standing questions. It's not just a snapshot of the last morsel of research; it's got to have perspective.

The desire and the necessity to write things like that increased. In 1996, 1997, we had some information on the evolution of limbs—the deep origin of limbs and some interesting comparative data with respect to vertebrate limbs. Neil Shubin, a paleontologist, invited me to write something with him and Cliff Tabin on the origin and evolution of limbs, and wow, three heads are better than one!

Then post-human genome project, there started to be a lot of chatter about human evolution. But some of the things being said, I felt, were not well grounded in what we already knew from model animals.

Gitschier: Like what?

Carroll: Too much anticipation that coding changes in proteins would explain a lot of our differences. Because, from the viewpoint of evolution of morphology, that was not what we were finding. The evolution of the human form—brains, bipedalism, neural wiring—I was motivated to tackle this. Can we anticipate what human evolution is all about, based on what we know about model organisms?

So I wrote a review article on that. That got me up to speed on hominid paleontology. I met paleontologists, read their papers. Hominid paleontology frames the issues. You've got to know the time scale of human evolution. At that point I had enough familiarity with a swath of material to tackle a book.

Gitschier: So how did that book [*Endless Forms Most Beautiful*] get off the ground?

Carroll: The trigger was that I was at a meeting, strolling the booths, and a Norton editor grabbed me. From their intel, they had heard that evo devo was something important. And they said they wanted to do something.

Gitschier: And this was at the same time you were thinking of doing a book.

Carroll: I was being asked to give some talks to general audiences about evolution of form. It's interesting to try to convey something in 50 minutes, but it was a vapor. How would the audience hold onto this? So a book would be a natural resource that they could have to follow through on some of this. I had a lot of warm-up for evo devo, so it was easy to get the riff going.

But I didn't know how to do a trade book—I didn't have an outline in my mind. With the live interest of Norton, I got serious. So that was the first step to entering this world, and it is a very different world!

Gitschier: In what sense?

Carroll: You have to learn some of its practices.

Gitschier: Like what?

Carroll: I did book tours each time. Regional NPR and National NPR, print interviews, public speaking, bookstore signings, doing Science Friday, giving a talk at a museum.

Gitschier: Did you enjoy that?

Carroll: Interesting experience.

Gitschier: So you are just a writing machine now.

Carroll: Now, it's psychotic. This is unsustainable—physically. Writing at the pace I have, putting out these two books. On the one hand, it is great for my soul. It's an interesting and creative challenge and very personally satisfying to hold yourself up and feel you have tackled some of that challenge.

And, as you get older in this business, and things aren't happening with your own pipetmen, it is nice to deploy a skill set and have some work to show for *yourself*. I'm sure it's made me a better scientist, because I think through things. I'm sure it's made me a better teacher, to be able to explain these things.

It's interesting too, working on evolution right now, because we have in this country the re-emergence of the dark ages!

So now, through book tours, I know writers at the daily newspapers, the commentators and the hosts. And I'm happy to be available when they have a question. I feel that part of my job is just to assist the media, because that's where people are getting a lot of their information.

Your sphere changes, and your sense of responsibility changes. You write these books, and you say, who reads them? Well—biology teachers! They rely on this to keep up with the science—I've worked with state and national teaching associations, my kids' school district, college board advanced placement test. What an interesting community that is!

You talk about outreach—this is what it's all about, and guess what, folks—it takes *time*!

I have just finished the third trade book and a spinout—a student book. Writing that, in the pure sense, was great. I was wading through the rich lore of natural history and some of the greatest people who ever lived—who wouldn't enjoy that? And I get to retell their stories in my own way but at the same time we've got some kick-tail research going on! And I have talks to give and journals to edit, so it's tricky.

Gitschier: I thought you said this writing business wasn't sustainable!

Carroll: This is *it*!

Gitschier: Can I quote you?

Carroll: Yes.

The student book is called "Into the Jungle"—subtitle: "Great Adventures in the Search for Evolution." The premise is that textbooks—let's speak in genetic terms—they are necessary but insufficient. They don't convey the process of how science really gets done, and they don't give you any sense of the personality of the individual and the human drama of discovery.

Gitschier: That's why I do interviews!

Carroll: Exactly why! You wouldn't want to study movies by reading a textbook on movies—you want to *see* the stories!

So I felt that one thing I could do to contribute to teaching evolutionary science in a better way was to change the format. The notion of the book is that if a student could sit down with stories, much as they would sit down with a book of short stories in English lit class, and enjoy the stories the same way they would enjoy fiction—the drama, the characters, the places—that they would have a different experience and that the science would come across by osmosis, not by pedagogical hammer.

The people who first went into the jungle in the search for the origins of species were really admirable people who did remarkable things: Wallace, who lost all his specimens in shipwreck and was in open sea for 10 days; Dubois, who decided to quit his medical career and went off to find "apeman" fossils in Indonesia and discovered *Homo erectus*! He threw the golden dart.

I feel if that a student curls up with these stories in 10 or 12 pages, they can't shake it off. It shows how serendipity plays a role, and how, even when you find something great, the world isn't always ready to recognize it.

Let me show you something [as he opens a large old book]. This is what I get to read when I research a story. In the early 1920s Roy Chapman Andrews went across the Gobi [desert] in search of ancient hominids—didn't find a one—but he went out in a fleet of Dodge cars with a camel caravan and this is the account of those expeditions. But instead, he found dinosaurs. These are the first dinosaur eggs.

Gitschier: How do you find all this stuff?

Carroll: I can't even tell you my process. I have a lot of help and a great library. So it's evolved. It's had its own little evolution.

Meeting a Fork in the Road

An Interview with Tom Cech

The past few decades have proven RNA to be a highly versatile molecule, with structural, regulatory, and even catalytic abilities, so much so that RNA is now considered life's primordial polymer. This interview explores the discovery that RNA can catalyze an enzymatic reaction on itself during its maturation. This process, known as RNA splicing, involves removal of an intervening sequence from a single continuous molecule of RNA and ligation of adjacent sequences to yield a functional RNA molecule.

Interviewed May 26, 2005

Published December 30, 2005

LONG BEFORE HIS METEORIC RISE TO A NOBEL PRIZE AT AGE 42 and to the presidency of the Howard Hughes Medical Institute (HHMI) a decade later, Tom Cech was just another guy working down the hall from me. In the mid 1970s, our lives intersected at the Massachusetts Institute of Technology (MIT) when he was a postdoc and I a graduate student. I recall a lanky midwesterner, modest and likeable, earnestly working on a project that didn't seem too exciting, or certainly didn't seem to be the stuff of scientific breakthrough, prizes, or fame.

I took advantage of our brief shared history to enlist Tom for an interview. We set up a video conference: Tom, nursing what appeared to be a green bottle of Perrier at the HHMI headquarters, and I, armed with a re-heated cup of coffee in the pristine confines of the HHMI conference center at the University of California San Francisco. The video transmission suffered from episodes of stutter and delay, with Tom's voice and image out of sync, disconcertingly causing Tom to look like a man on a lunar mission in an Apollo spacecraft, rather than safely ensconced in Chevy Chase.

We got the ball rolling with a bit of reminiscing about our experience 30 years ago at MIT.

Gitschier: When I came to MIT in September of 1975, I recall being thrust into an exhilarating atmosphere. The first seminar of the year, called together hastily, was by Phil Sharp, who had just returned from Cold Spring Harbor to describe RNA splicing for the first time. One month later, champagne was flowing in the hallways because David Baltimore was awarded a Nobel Prize. It was something!

Cech: I had the same experience. We were in the old-fashioned biology building, 16–56, but it was a very exciting place. My benchmate [in Mary Lou Pardue's lab] was Joan Ruderman, and Matt Scott and Al Spradling were in the lab at the same time. It was a very small lab, but half the people ended up getting elected to the National Academy of Sciences. There was a very high level of scholarly interest, good critical thinking, and a lot of excitement about science.

Gitschier: I recall you studying psoralen cross-linking. What were you really up to?

Cech: I was looking for a way to freeze nucleic acid structures in vivo, and then to pull them out without rearranging them. I was particularly interested in alternative secondary structures of DNA, cruciform structures, and I was doing almost exclusively electron microscopy. I did that for a long time! I was trained in the Cal Tech/Norm Davidson/Phil Sharp/Ron Davis tradition of quantitative DNA electron microscopy, so you always had to measure hundreds of molecules before you believed what you were seeing in the EM [electron microscope].

Most DNA molecules look very boring because they are just a double helix. But if it's branched, or if it's a replication form or a Holliday junction or some alternative form, it [psoralen cross-linking] can be a useful tool. But then John Hearst's lab found that information about where the nucleosomes were located on the DNA was also preserved because the regions that were in the linkers between nucleosomes were much more accessible to cross-linking than the regions wrapped around the histones.

I built a zapper. I bought a high-intensity mercury vapor lamp, the kind used in streetlights, and I set it all up myself! I was amazed that no one got killed because I never had any training. I built this thing with the shield and the switch and the temperature-controlled reaction chamber. You don't realize how bright and hot these streetlights are till you get one a foot away. There is an immense flux of light coming out of there. And that's the wavelength of radiation that activates the psoralen cross-linking. That thing may still be there at MIT.

Gitschier: So you moved to Boulder, left the cross-linking, and moved on to ribosomal RNA genes.

Cech: The main thing I decided was to do something completely different when I started my faculty position.

Gitschier: That seems unusual; most people capitalize on the progress they've made as a postdoc.

Cech: I think it was a little less unusual then than now. I wanted to look at the chromatin structure of a transcription unit. People could clone DNA by then, but they couldn't clone DNA with its natural histone and nonhistone proteins associated with it in the right place. Tetrahymena had 10,000 identical copies of the ribosomal gene. I thought that one could isolate the gene in its chromatin state and look for regulation of transcription. It was a fine idea—it never worked.

Gitschier: But fortunately, something else worked!

Cech: But fortunately, the gene had an intron, which it wasn't even supposed to have! As you mentioned, this was right after Phil Sharp and Cold Spring Harbor had discovered RNA splicing. By the time I was an assistant professor, there were 100 examples of eukaryotic introns, but no information about splicing mechanisms. In fact, people were just starting to come to grips with the fact that it was splicing rather than some kind of transcriptional jumping.

Gitschier: I remember that debate.

Cech: There were several papers, including one by Shirley Tilghman and Phil Leder, showing that there was a precursor RNA that was then processed into spliced RNA, so it looked like splicing was operating at the RNA level. But there was only one lab, John Abelson's in Southern California, studying yeast splicing, which had much of a handle on splicing in vitro.

So when we found that the intron in our ribosomal genes was being spliced, there was a fork in the road. We could either say, "We're not funded to do splicing research by the NIH [National Institutes of Health]; we're funded to do transcriptional regulation, so let's just report that intron and leave that for someone else to follow up," or we could follow this splicing angle. We tried to do both for a while, but of course the splicing became more interesting.

Gitschier: Tell me more about that discovery.

Cech: It was during a period of about a year when I was insisting that there had to be a protein enzyme stuck to the RNA that was doing the splicing, and we were trying to shake it off so that we could get back to studying the splicing protein. Boiling the RNA, adding ionic detergent, boiling in the presence of detergents, adding proteases—all had no effect. We were getting more and more desperate!

As time went on, it seemed less and less likely that there could be a protein, but if we announced that the RNA was self-splicing, no one was going to believe us! So we then turned to recombinant DNA, which we didn't know anything about, and we made an artificial gene with a promoter and purified it in *E. coli* [*Escherichia Coli*]. When RNA transcribed in vitro from the artificial gene underwent splicing, then we were convinced.

During that year, there was a slow transition from thinking there was a protein to thinking there couldn't be one. By the time we did the final experiment, it had to work without the protein because there was no backup plan.

Gitschier: Have you had any other "eureka" moments?

Cech: We've been fortunate to have a couple more, but they weren't as exciting to me because I was doing those [splicing] experiments with my own hands, together with Arthur Zaug and Paula Grabowski. That is a special kind of "eureka" moment.

Since then, we've had a couple of big moments in the lab. One was finding the catalytic subunit of telomerase, something that had been predicted by Liz Blackburn and Carol Greider as early as 1985, when they described telomerase and found the RNA subunit. But ten years had gone by, and the whole world and a couple of biotech companies were searching for the catalytic subunit, when a postdoc in my lab, Joachim Lingner, purified the catalytic subunit from euplotes. We called it TERT telomerase for "telomerase reverse transcriptase."

And in 2001, Peter Baumann found a protein that caps off the very ends of human chromosomes; we called it POT1 for "protection of telomeres." We found it first in fission yeast, and did the genetics there.

The human genome project is great—we do all this work in simple organisms, and if you believe in evolution, you believe these things are probably widespread and can be found in the human database. That was another very exciting moment, again, because these telomere-capping proteins had been found 20 years earlier in ciliates, and people had been wondering whether the ciliates were special or whether all eukaryotes would have them.

Gitschier: Were you disappointed that these weren't RNA molecules?

Cech: Well, it was exciting enough that telomerase was half RNA.

Gitschier: Can you speak about how our whole perspective has changed in the last 20 years since the discovery of RNA as a catalytic moiety?

Cech: There are so many areas in which the RNA science has provided excitement. The whole nature of the ribosome, for example, and the work in Harry Noller's lab at UCSC [University of California Santa Cruz]—where his experiments indicated very strongly that RNA in the ribosome was doing more than just scaffolding key proteins, and that the RNA was the important entity for peptidyl transfer. That work came to its culmination when Tom Steitz and Peter Moore determined the crystal structure of the large subunit of the ribosome, showing that at the peptidyl transfer center, there was only RNA.

But look back at Jacob and Monod's early thinking that the lac repressor might be an RNA molecule. Once the lac repressor was proven to be a protein, people got all focused on gene expression being regulated by proteins. But if they had been looking in bacillus, rather than in *E. coli*, they would have found riboswitches built into transcripts that bind small-molecule metabolites and control transcriptional termination or translation. So the old Jacob/Monod model was actually right, but for a different branch of bacteria.

And of course, you can't even turn around without seeing RNAi [RNA interference] and siRNA [small interfering RNA] all over the place. So chances are there are still a lot of RNA-level functions to be discovered in complex genomes.

Gitschier: In an interview, you said you were a prolific writer as a younger man. What kind of writing did you enjoy, and do you still write?

Cech: I don't know that I ever enjoyed writing—writing is painful!

I started with a fourth grade teacher who made us write essays till our hands felt like they would fall off. I did a lot of writing at Grinnell College—humanities courses, great books courses, and a constitutional history course. I think learning to write a good argument and support a point of view in the humanities really helps your scientific writing. Scientific writing is particularly difficult, and you need all the help you can get. I've never written fiction or poetry or anything really interesting.

Gitschier: Do you ever think about attempting that?

Cech: I do think often about writing a book about my own experience in science. I like these books that some of my colleagues have written because they humanize the process, and they show how twisted the path to scientific knowledge is. I think there is so little understanding in the general public about the life of a scientist or the scientific process, and especially about the intense skepticism that we bring to everything we hear about or to even the discoveries in our own lab. I think if more of the problems in politics, governmental policy, and international affairs were approached with more of a scientific attitude, we would all be better off.

Gitschier: I would agree with you, and I would encourage you to follow through on that. What have been some of your favorites?

Cech: I re-read Watson's *Double Helix* recently. It's really well written: the excitement, the uncertainty, the despair and disappointment, and the exhilaration. That's probably the best example of a scientist giving an account of a portion of his own work.

Gitschier: You love doing research, and you're obviously so good at it, so why did you take the job of president of the HHMI?

Cech: After the Nobel Prize, I had a great decade of additional work and a big lab of 25 or more students and postdocs, who are now professors at good places around the country. And I did a lot of teaching; for instance, I insisted on teaching general chemistry for freshmen for six years.

But I felt I was making an impact on only a local level, and I had a need to make an impact more nationally. So I started to serve on some scientific advisory boards. It was useful; however, I'd go there and have a great interaction, but then I'd leave and the people [I was advising] were still stuck there. If you really want to nurture something, you have to be there when the initial excitement subsides.

So I thought I would consider moving into a leadership position, but promised my family I would wait till my daughters finished high school. Then the Hughes position came along and that seemed so special, and one that doesn't come along very often. It was the idea of having an impact on the educational programs that was the big driver for me. The investigator program was so well run that there wasn't so much of a challenge.

Gitschier: Max Cowan [former vice president of the HHMI] told me that you were one of his top choices for the HHMI presidency, specifically because of your keen commitment to teaching.

Cech: Is that right? Well, thank you! He never shared that with me, although we always had a very good relationship.

We've evaluated all the educational programs; we've discontinued a number of them, started new ones, and massaged others. The role of a nonprofit is to do something different from the federal government, including things that have some risk, because you want to try something that can have a huge impact. But then you have to have the discipline and strength to discontinue things if they aren't successful.

Every time we've discontinued a program, there is a firestorm of outrage in the community, and I get accosted personally and get a lot of angry letters, but that's part of being in leadership—to evaluate and get good advice from the best people you can find, including the trustees, and then to just do it.

Gitschier: So, how do you like the job?

Cech: There are large parts that I find to be enjoyable, and there are also significant parts that I find important to be done well but which there's not much enjoyment in. You know, it's really wonderful being a university professor!

Gitschier: I know!

Cech: There is tremendous amount of freedom, and people don't yell at you very much either!

In this job, you make a lot of people happy and a lot of people very unhappy, and the people who you make unhappy are sometimes gracious and understanding, but sometimes they'll never forgive you. But you have a process, and you have to take the scientific review board advice extremely seriously. And it's also really important for there to be turnover.

Gitschier: Now that you are at the helm of the HHMI enterprise, you have a special vantage point overlooking biomedical research. What do you find the most interesting?

Cech: The most eye-opening for me has been the neurosciences, probably because it's the area where I was the most ignorant, but also because it's extremely exciting to be probing the cellular and molecular basis of behavior, sensory perception, and memory, and moving toward understanding consciousness and cognition. I find the neuroscience talks to be the most riveting.

The other area, also very far from my upbringing, that I find to be so gratifying and so important is translational medical research. People doing really high-quality basic research stimulated by what they have seen in the clinic. It is really exciting when one of the investigators makes a mouse model that recapitulates a human disease, treats the mouse, and gets an effect, and now is treating patients and gets remediation of the disease.

Gitschier: So if you were going to start a postdoc today, what would you choose?

Cech: I'd probably still be a boring chemist because I personally love thinking about the molecules. I do want to be thinking about molecules that are important for human health.

I think you have to do what your passion is. My passion is to think about nucleic acids folding up and interacting with themselves, interacting with small molecules, and interacting with proteins, and also to think about biochemical reactions, about how their specificity and rate enhancement is generated. In ten years, the stuff we're doing will probably be totally unfundable!

Gitschier: How long do you think you'll be staying with the HHMI?

Cech: I serve at the pleasure of the trustees, and so far they've been supportive. Things are still exciting. It's been very important for me to have been able to continue some research because it uses a completely different set of neurons than administrative work.

There is a search for absolute truth in research. You never get there—but there are criteria by which you judge how close you are. You're always criticizing yourself and criticizing your colleagues, and they're criticizing you. And there is a test, very often, that you can do to decide who's right.

But in administrative work, it's all about empowering people, and there is never any absolute truth, and you can never fool yourself into thinking you've made all the right decisions. They are completely different jobs, and I enjoy them both.

There is perceived value to Hughes to have the leadership actively engaged in science. The real question is what will happen to my 30-year NIH grant coming up for renewal next year.

Twenty Paces from History

An Interview with Soraya de Chadarevian

In the history of scientific revolutions, the prominence of Cambridge, England, is perhaps unequaled. This interview considers how a historian of science looks back over a period of great progress, in this case the birth of molecular biology in the Cavendish laboratory after World War II.

Interviewed June 26, 2006

Published September 29, 2006

WHAT IS IT ABOUT CAMBRIDGE THAT SO DELIGHTS ME? Could it be the soothing rhythm of punts and poles upon the Cam? The fan vaulting of King's chapel, rendered even more breathtaking by the dappling of light through stained glass? The rickety Dutch bikes spilling into narrow streets as residents sensibly eschew motor power? Or its lure as birthplace of molecular biology, a modern chapter in the lineage of scientific inquiry nucleated here by the likes of Newton and Darwin? The answer is all of the above.

Each time I visit Cambridge, I succumb to the magnetism of Free School Lane, and like an iron filing I'm pulled deeper and deeper into the interior courtyard of the Cavendish labs. It was here, on the "first floor" of the Austin wing, where Watson joined Crick and the structure of DNA was deduced. Last year, I prevailed upon an affable ginger-haired woman, who worked in a hut nearby, to direct me to this hallowed space. She seemed to take delight in my quest and pointed out a window in the large yellow-brick structure. "There," she said, her friend worked in the very room where the model was built. I looked up and knew she was right.

To learn more, I turned to the Department of History and Philosophy of Science, adjacent to the Cavendish, and was referred to the book *Designs for Life*, a history of molecular biology in Cambridge, written by one of their faculty, Soraya de Chadarevian. I darted over to the Cambridge University Press bookshop, picked up a copy, and dove into it on my way back to the United States. One year later I took a sabbatical at the Wellcome Trust Sanger Institute, grandchild of the noble Cavendish, and I had the opportunity to visit the author in her small office overlooking history. What, I wondered, was it like to immerse oneself in that remarkable period?

Gitschier: How did you happen to come to Cambridge?

de Chadarevian: I came here fifteen years ago on a three-year Wellcome fellowship to work on a project on the history of molecular biology in Cambridge.

The Wellcome Trust funded several Units for the History of Medicine, and one Unit was here in Cambridge, under one roof with the Department of History and Philosophy of Science. The Unit had decided to launch this topic as a project. They were looking for a researcher to do the main work on the project. They hired me.

Gitschier: What was your background?

de Chadarevian: My first degree was in biology and then I did my PhD in philosophy. I had just done a post-doc in history of science in Berlin. I was working on a

19th century project, and I liked that a lot. But it was suggested to me that if I wanted to stay in the history of science, I should consider moving to the 20th century so that I could use my science skills.

This is always an issue: do you need to have a science background when you work in the history of science, or don't you? I think either is possible, but for me it has been a big advantage that I had science background.

First, it was easier that I didn't have to learn the science, but also I personally think it is difficult to write the kind of things I write without having been in the lab, without having an idea of what it means to do science. And when you interview scientists, they talk to you in a very different way if they know that you understand what they say. I think they respect you more. I think it has been crucial that I have a science degree.

Gitschier: Where did you study?

de Chadarevian: I did a Diplom in Biology in Germany. I grew up in Italy, but I did most of my studies in Germany.

Gitschier: But your name is not Italian.

de Chadarevian: My first name is Arabic, the last name is Armenian. But my mother was German, so I went to the German School in Rome and then won a fellowship to study at the University of Freiburg. The Diplom is a five-year course, which includes an extra year of experimental work, which I did in Bologna.

Gitschier: What kind of research did you do?

de Chadarevian: I worked on bioenergetics, on the mechanism of ATP formation in the cell. A group at the University of Warwick in the United Kingdom was promoting a chemical-coupling hypothesis, in contrast to the electrochemical hypothesis proposed by Mitchell. In the chemical-coupling hypothesis, there was one essential ingredient—lipoic acid—so we worked on bacterial mutants that were lipoic-acid deficient.

It was fantastic being a graduate student, having a project that was your own. And it was really a clear-cut question, a textbook question. And, actually, my experience of how science works was very important for my future work on the history of science. For instance, we got our results very quickly and clearly, but we didn't believe them, because they contradicted the results of this other group in Warwick.

A simple experiment became excessively complicated, because the group from the other lab suggested that lipoic acid could linger in trace amounts and that we were not working cleanly enough. Or, perhaps, that the water in Bologna was different from the water in Warwick. A research assistant came down to

Bologna to work with me, because the idea was that we were doing something different, something that couldn't really be figured out from the recipes: the idea of tacit knowledge that goes into doing an experiment but not written into the methods section of papers. It is very relevant for the approaches in history of science, which I was to learn later, but here I was living it!

Still, we couldn't reproduce the results of the competing group. In the end, it turned out that theirs was a case of fraud! Then there was another group, Canadian, that got results similar to ours. At that point we very quickly published our results.

For me, this was a lesson in how science works and how knowledge is made. Sociologists have called it the "experimenter's regress." It means that when you do an experiment and get some unexpected results, you don't know whether to trust the results or to suspect that something is wrong with the way the experiment was done. And you can't decide in advance.

During that time, I became interested in thinking about how science works, how knowledge is produced, and about the place of science in society more generally. At that time this meant going into philosophy, because history of science in Germany at the time was much more of the antiquarian sort—who invented what and when—and I thought that was boring.

After receiving my PhD, I learned about a German postdoctoral fellowship program aimed at introducing new approaches to the history of science, and I became one of the first generation of fellows in that program in Berlin.

Gitschier: So what was this new way of thinking about science?

de Chadarevian: It came from many directions. The general idea was to look at science as a practice, as a culture. It involved anthropologists moving to the laboratories and looking at scientists as a foreign tribe. They would ask: What are scientists after? How do they talk to each other? And what do they actually do?

It also came from the sociology of science, and from taking seriously the material culture of science, not looking at science just as a system of theories, but as a system of practices. And making links to other cultural practices.

During the 1980s, the history of science became really interesting, because there were people coming from many different disciplines and backgrounds and working together in new ways, examining what scientists did in the lab and beyond. There was a very special dynamic that became extremely productive.

We are talking about the history of this discipline. These new approaches came out of a particular historical moment, when new questions were raised about the role of science in society and about the role of the expert. Many of the people who started to question this were scientists themselves. This had started in the late 1960s with the critique of the role of science and technology in the Vietnam War.

Gitschier: Let's now talk about your book.

de Chadarevian: The project changed dramatically in the course of the time I worked on it. I worked on it for a long time! The gestation period for the research and for writing the book and getting it out was nearly a decade, which many people think is what you need for a historical project. Other things came out, too, including two edited volumes, an exhibition, and a series of articles, all of which were part of the big project.

Gitschier: How did the project evolve?

de Chadarevian: At the beginning it wasn't very well defined at all. The organizers had the feeling that Watson and Crick had been here, so somehow molecular biology started here, and they wanted to see how it established itself as a discipline.

What I did was to place these developments into the broader historical context of where science and scientists were standing at the time. People might be surprised that the first chapter in the book talks about the mobilization of scientists in the war. [Indeed, the subtitle is *Molecular Biology after World War II*]. But I can tell you how I got to that.

There was a small steering board for this project. One of the people on the board was John Kendrew, and he was one of the first people I interviewed. It was in talking to him, and later to others, that the experience of being a scientist in the war was so dominant and so important in their telling of their lives as scientists, that it just struck me that this had to be part of the story I was going to tell. And it struck me particularly in coming from Germany. German scientists would have done everything to avoid that subject! The experiences and skills scientists gained working on war-related projects became important for their careers and for their research after the war.

Also pivotal was that funding for science changed dramatically after the war, as a result of the war. Funding for science projects escalated during the war, and scientists made the case that this funding had to be kept up after the war. They convinced the politicians. In Britain, this became a very strong topic. The scientists had contributed to winning the war—see radar, see the atomic bomb—and they would help the country win the battle of peace, in the rhetoric of the time. Molecular biology didn't exist at the time, but "biophysics" attracted much funding. It included the use of physical technologies, often developed in weapon-related research, for biological and medical purposes.

Today there is a lot of research done on the science of the 1940s and 1950s. There is always a lag time because historians need to wait for the opening of archives, and, in this particular case, they also needed to wait for the end of the Cold War. People are now working, for example, on the history of radioisotopes and how their use in biology and medicine was actively promoted by the Atomic Energy Commission in the US—the peaceful side, the good side, of physics, something that could somehow redeem the destructiveness of the bomb. People were speaking in these terms, suggesting that the medical uses of radiation could save more lives than were lost in Hiroshima and Nagasaki.

Postwar, science funds were expanding and there were opportunities for new research. People were coming back to research after long interruptions in their scientific careers. And they were prepared to move into completely new fields. Instead of choosing a topic that was more or less suggested by their supervisor, they decided to do something they were really interested in, since they had to start from scratch anyway. For example, Francis Crick in his autobiography speaks in such terms. After the war, he wasn't keen on continuing his PhD topic, which he considered "extremely dull," in his own words. Luckily, he said, the war had destroyed his experimental setup, and so after the war he couldn't go back to it.

When I started my project, I had made one decision: people had written so much about Watson and Crick already, even before the 2003 50th anniversary, that I didn't want to focus on that. I wanted to look further afield.

But eventually I realized that the double helix discovery so dominated the history of the field that I just had to engage with it. Writing the history of a field also means reflecting on the way that history has been written before and on how history is used, for instance, by scientists to construct a particular picture of their science.

In the case of Watson and Crick, the question is why has that story become so important? You have to understand its place and meaning before you can see what other stories you might be able to tell.

Gitschier: Well, Watson himself was such a great spokesman for the story.

de Chadarevian: Of course. But one needs to understand what kind of a story it is. It is not history. Actually, Watson himself called his book *A Personal Account of the Discovery of the Structure of DNA*. He didn't even claim it was history.

Gitschier: How do you define history, then?

de Chadarevian: History is done by looking at events with some distance, by studying the papers and putting things in context. The same is true of general history.

Gitschier: But how do you know that papers, and in fact this is where my argument lies for oral history, actually reflect what really happened at the time?

de Chadarevian: I am talking about papers in a broad sense. The papers kept in archives, including personal archives, institutional papers, and correspondence of the time. Of course papers always need to be placed and understood in the context in which they were produced. They don't tell the story "as it was."

Gitschier: What are we going to do in the future with everything these days being e-mail?

de Chadarevian: That is a big problem. Archivists are trying to think about that. For instance, John Sulston has kept all of his e-mails and they form part of his archive that will be deposited with the Wellcome Library.

Gitschier: Yes, I read Sulston's book [*The Common Thread*], and I was amazed that he cited e-mails from five and ten years ago.

de Chadarevian: Sulston used e-mail systems early on. They were crucial for communicating in the collaborative networks he was working in, and he was aware of the importance of that. There is an extremely interesting history to be written about this early electronic communication between scientists and how that affected the kind of science that could be done. So that is the kind of question we are interested in.

The human genome project poses other questions as well. For instance, where is the discovery in that kind of research? This has been the problem in high-energy physics all along, where hundreds of people are involved in single experiments. Who is the author? Who gets the Nobel prize? The whole credit system, as well as the notion of discovery, is called into question.

Linking the human genome sequencing project to the Watson and Crick story, as is routinely done, means also projecting a heroic discovery story on the more tedious work of sequencing. People have argued that this could help make the sequencing project look more exciting.

Gitschier: What about the use of interviews in your research?

de Chadarevian: Interviews are extremely useful, but as historical sources you can't take them at face value. I have actually written about that—how to use interviews for writing the history of science, if you are interested!

Interviews happen usually in a dialog, and the account is really a co-production of these two individuals. A question is asked and a response given. So the response is directed in a certain way and it's always a reflection on past events, not a simple description of what happened in the past.

And then there is the way memory works. People get things completely wrong retrospectively. Sometimes there are conscious alterations, like cutting out details the interviewee prefers not to talk about. But memory also works selectively. And we forget things. A story, told many times, can become more real than the real thing. That is why a letter written at the time is generally more reliable a source than a memory of 30 years ago.

Gitschier: But scientists keep a laboratory record of some kind.

de Chadarevian: Sometimes they are actually very surprised by what they find there! Because they have become convinced of the development of a particular series of events, but when they trace it back, they find it might not have happened the way they remember it.

Interviews for research purposes are very time-intensive. You have to work a lot in advance to be able to ask questions that will move beyond what people have written before. And then, after the interview, it's extremely time-consuming to transcribe it and then to check all the information. So I have done some but not too many interviews. I mainly view interviews as leads to historical research, to get to papers, to check interpretations. It's not the end product for me. It is just one part of the mosaic, one source among many others.

Gitschier: What about the archives?

de Chadarevian: There are rules that govern the access to archives. In Britain, for instance, there is the 30-year rule for all government papers. Personal files are closed for 50 years beyond the death of the person. That's why there is always a lag time before historians take up a subject. And that's why many people think you can't do the history of the 1980s yet.

Kendrew featured quite strongly in my book because he deposited his very substantive collection of papers in the Bodleian [library at Oxford University] when he was still alive, and he collaborated in the production of the catalogue, which alone has about 900 pages! I needed his personal permission to get to the papers, and at the beginning he was rather restrictive. But with time I gained his trust, and by the end I had free access to practically everything.

These are the obvious things—personal papers and institutional papers—but then you have to look around. For instance, Max Perutz got a grant from the Rockefeller Foundation, so you go to their archives and see what's there. They have rich archives, including the diaries of their officers who regularly visited different laboratories and provided long descriptions of what they saw at different times.

Or you start being interested in the models of proteins and DNA and in the television programs about them. There is a wonderful BBC archive with correspondence between the producers and the scientists. You can trace the initiative for a specific program, and how the project changed along the way. What did scientists think about their collaboration? Were they fearful of the image they projected? This is science in the public realm that is moving far beyond science in the published scientific papers.

Gitschier: Who were you thinking of as your audience?

de Chadarevian: This is a big issue in the history of science. Everyone feels that writing only for those five other historians of science who work in a similar field is not enough. Also, history of science is becoming so much a topic for popular writing. People speak about the "Sobel effect," after Dava Sobel, the author of *Longitude*, which had such a big impact.

Gitschier: That's a great book. I actually went to Greenwich a few weeks ago to see the Harrison clocks!

de Chadarevian: Historians also want to reach broader audiences. But you have to make sure the scholarship is not lost. Some people feel you have to do both, writing academic books and articles and more popular books.

The scientists are an audience I think we have to be especially concerned with because it is a huge market. Also, if we want to make an impact in the way science is perceived, then if the scientists themselves are not interested in what we write, there is a serious problem.

Gitschier: To devote ten years of your work life to others' work, it strikes me that you would have to be drawn in to the characters. Perhaps you became quite fond of them.

de Chadarevian: I kept a distance, by design.

Gitschier: I don't think, then, that I could be a historian of science!

de Chadarevian: It doesn't mean that you can't be enthusiastic about it. I say this in the preface of my book. Every time I come to my office, I pass by the old Cavendish laboratory, and in the summer especially there is always a group of tourists. As I pass by, I wait a bit to hear "Watson and Crick," and so on, and that has become part of this cultural heritage. It was special writing that story here.

Gitschier: I confess, I think I'm particularly drawn to the double helix discovery because it occurred during my birth year.

de Chadarevian: And I was born on the very day that famous photo of Watson and Crick with the double helix was taken!

We smiled at each other. And like a co-conspirator, she offered to show me the room shared by Watson and Crick when they toyed with their models. We went through the Cavendish portal, crossed a small courtyard, and turned right to enter the Austin wing, where the Department of Materials Science and Metallurgy is now housed. We slid up a set of stairs, hit the buzzer, and were admitted to a long corridor. There, three doors down on the left, was room 103, where a solitary woman sat, her back facing us. We matched up the verticals bands on the south wall with those in the backdrop of the photograph and convinced ourselves that this was the spot. The attraction was so strong, I found it hard to pull myself away.

Stable in a Genome of Instability

An Interview with Evan Eichler

The genome of any organism has an architecture, generated by duplications, deletions, and rearrangements of DNA, that reflects its evolution. This interview addresses three interesting consequences of this process: first, that any two individuals can vary significantly in their genetic architecture; second, that sometimes in attempting to heal itself of a DNA duplication, deleterious deletions can emerge; and third, that comparison of the genomic architectures in different species can provide insights into the evolution of our own genome.

Interviewed January 30, 2008

Published July 25, 2008

W<small>E LIKE TO THINK THAT OUR GENOME IS ROCK-SOLID</small>, that it is dependable, there for us when we need it. The truth is far from that. By fits and starts, our species' collective genome is undulating, reshaping itself with eruptions of genomic lava and clashes of sequence tectonics, at once both marvelous and unsettling. We are unaware of this tumult within us until we are confronted with disease in ourselves, our friends, or our family.

Evan Eichler is a man obsessed with this process, and to speak with him is a study in contrasts. An unassuming Canadian, Eichler is a student of genomic architecture, the arrangement of sequences in our genome, and their evolution. Eichler grew up on a farm in Manitoba, married his college sweetheart, and now lives together with her and their four children in the mountains east of Seattle. As we walked up the hill to my office during his recent visit to UCSF, he talked about being an early riser, taking his son to band practice before school, and then driving the 30 miles to work in his Toyota. Eichler is a man bristling with excitement for his discoveries, but holding it in check by a tradition of modesty. He has consistently followed his own path, chosen career opportunities that were dictated not by politics or peer pressure but rather by what feels like a good fit for him.

Our conversation ranged from tiny triplet repeats to large and complicated duplications, some of which harbor genes of uniquely human import, and the process of their discoveries.

Gitschier: Your thesis advisor, David Nelson, told me that when you came to Baylor [College of Medicine] as a first-year graduate student, you already knew you wanted to study genome evolution.

Eichler: That's true.

Gitschier: How did you know that?

Eichler: I was at the University of Saskatchewan [as an undergraduate] and I was in Biology. They didn't have a genetics program, but I knew even before I went there that I wanted to do genetics.

Gitschier: And how did you know *that*?

Eichler: It started out in grade 9 or 10. My family grew up on a farm, and we started to raise angora rabbits for the purpose of their wool. My mother was one of those folks for whom everything had to be done naturally. So we had to pull the wool, we couldn't clip the wool. (It's OK—the rabbits are fine with it!) She spun her own wool. And she didn't believe in dyeing the wool.

She said to me, "I want different colors of wool, but I don't want to use dyes." And in grade 9, I learned how to use the Punnett Square to keep track of the five gene coat color system in rabbits. I got a little textbook, and I started breeding these rabbits. I joke that that was the only time I ever did classical genetics!

I did those experiments, and within a couple of generations, I got all the colors that she wanted. I could breed them true. My mother was so impressed! It's amazing what you can do!

At that point, I decided that genetics was what I wanted to do. And by the high school years, after reading stuff, I realized I wanted to do human genetics.

So my father looked into a number of different universities to check out the genetics programs, but I ended up settling on a place where there was no genetics, because it was close to home.

I wanted to take more molecular courses, but I ended up taking more ecology, evolution, and anthropology, because it was part of the curriculum. They [the faculty there] didn't believe in what they called "reductionist" biology.

Gitschier: But that served you so well in the long run!

Eichler: This is, I think, where my interest in evolution [was] sparked. And when I finished my Bachelor's degree, one of my professors said, "If you want to do human genetics, you have to get an MD."

So I thought about that, and I took a year off. I got a fellowship to study in Munich, at a veterinary research institute. And there I got real exposure to research, and that's where I applied to different [graduate] schools—Sick Children's in Toronto, Hopkins, Yale, and Baylor. And got invitations from those schools but eventually went to Baylor. I thought the research there was comparable to that at the big names, and I thought the people like David and Phil Soriano, who had interviewed me on the phone, had this folksy feel—really down to earth, but very high energy and, obviously, top quality.

My uncles gave me a hard time: "You had a chance to go to an Ivy League school and you're going to *Texas*!" But at that time, Tom Caskey had such a great enterprise there. He had such great taste and recruited such impressive faculty. I was so happy there.

Gitschier: And you went, without even looking at it?

Eichler: I hadn't seen the city, and if I *had*, maybe I would have changed my mind! I went there. I wasn't married yet, so I had to fly back [to Canada], marry my wife in the middle of midterms and bring her down, and she said "I cannot believe you have moved me from Canada to *here*." She hated it for the first six months, but eventually she grew to like the city and the Medical Center.

I was extremely lucky to find David Nelson as my mentor. There was an instantaneous click—a chemistry. He gave me complete freedom, but he was an academic rock. When I came up with ideas, he would quickly find where the flaws were and then allow me to go on and pursue them. I was one of those strange students who actually wrote two qualifying exams because I couldn't decide what I wanted to do. My committee told me to focus, but David said, "Eh, do what you want," shrugging his shoulders.

Gitschier: I can just hear him saying that!

Eichler: And I loved this whole Fragile X thing—the idea of a mutation being dynamic, and a premutation state. The anticipation phenomenon [that disease risk increases in subsequent generations, now known to be due to triplet repeat expansion] had been rejected by a lot of mainstream geneticists ten years before—they thought it was just ascertainment bias. And then to have it all resolved by Ying-Hui Fu and David, to be there at that moment when those *Cell* papers were coming out! When I came to that lab, from early on, I was interested in studying that process from an evolutionary perspective.

When I think about duplications, I think about them *exactly* along those same lines—as a dynamic mutational process. Instead of slippage of triplet repeats, it is non-allelic homologous recombination. These regions, unlike most of the genome, break all the rules. They can have very accelerated rates, and then pause, if there is selection, either positive or negative. They beat to their own tempo.

I started working on the mechanism of the instability. Why do triplet repeats expand at all? From Ying-Hui's sequencing work, we knew there were AGG interruptions in the CGG repeats. So working with David and Steve Warren, we came up with a model that a loss of AGG interruptions would predispose alleles to change. We showed that alleles that lacked the AGG interruptions moved toward premutation and disease state much more quickly within the human population. Some populations, such as Tunisian Jews, had a disproportionately large number of uninterrupted alleles, and in these same populations, Fragile X syndrome was much higher. So what mattered [in promoting instability] was a pure tract of CGG rather than the total number of repeats.

Gitschier: Other primates don't have fragile sites, do they?

Eichler: Not that we ever have observed, and they also have many more interruptions in the CGG sequences, and different types of interruptions in different species. And you never see the amplification and the fragile site.

All microsatellite lengths in other species are shorter on average than in humans. Even the polyglutamine-coding tracts. It's almost as if the human species has been sloppy to allow these types of track lengths to increase, unless they have some kind of benefit.

At the end of my PhD, I got a side project going—to map the Emery muscular dystrophy gene. So I started mapping cosmid clones in the Xq28 region, and, lo and behold, as I was walking across that region I got some unusual results—cosmids that should have come from the X-chromosome, but hybridized by FISH to multiple locations. One clone had the creatine transporter locus, and it hybridized clearly to both Xq28 and 16p11.2. And another one—the adrenal leukodystrophy locus—it hybridized to four locations in addition to Xq28.

That's when I started the idea of looking at duplications and copy number variation, in 1996, 1997.

Gitschier: At the time, were you really thinking about copy number variation *within* species, or just about the evolution of segmental duplications?

Eichler: At that time I would have been thinking about segmental duplications and copy number variations *between* species. But it was shortly thereafter—1997, 1998—that copy number variations *within* a species became apparent.

There were pericentromeric duplications, proximal to the Pradi-Willi region on 15q, published by Marc Lalande, and another paper by John Barber reporting larger 16p11 copy-number variations. Both of these papers showed copy number variation in large segmental duplications in normal individuals, and most of this variation was thought to be evolutionarily quite young, less than 10,000 years. Barb Trask had shown in 1998 that the subtelomeric regions had dramatic structural variation between species and within humans. And all of this was *way* preceding any of the hype in 2004 about copy number variation.

Gitschier: I know what you mean. Suddenly there appeared an acronym "CNV" [copy number variation] for something that had been known about for quite a while.

Eichler: One of the problems with genomics, in particular, is that the collective memory seems to be about five minutes to twelve [o'clock]. It's where people are at the last three or four minutes before the bell tolls that seems to matter and they forget everything that went before. It's a little bit frustrating, but I imagine everybody feels this at some point. It's not that these ideas appear out of nowhere. It's not a vacuum and suddenly a light goes on one day! They're built upon many studies over many years.

So between 1996 and 1998, we were already thinking about copy number variation, but we wanted first to understand the organization [of duplications] in humans, and then understand the difference between humans and other primates, and then focus again on humans, distinguishing normal and disease-causing variation.

Gitschier: You must have great computational skills to do this kind of work.

Eichler: I'm not a programmer, and I never took a single class in it. But David was a big fan of UNIX. I never was afraid of moving big data sets around.

What happened to me was that we had done these anecdotal studies looking at duplications, repeats, variations between species, and then a couple of things happened.

First, I took my faculty position at Case Western Reserve University. Hunt [Willard] and Aravinda [Chakravarti] recruited me. David and I had been working on a paper with Aravinda on Fragile X haplotype analysis. He heard I was job hunting and said, "Why don't you come and look over here?" I thought, "That is a great place!" Hunt had the chromosome structure part, Aravinda was doing the human genetics disease angle, Rob Nicholls doing the Pradi-Willi/Angelman work. It felt like a natural fit with that whole faculty. And Cleveland is great! It was cheap. The people there seemed so down to earth. It was just a perfect fit for me and my research. So I moved there in 1997.

Gitschier: So now, let's talk about how you moved to the whole genome problem, which I assume is the second "thing" that happened.

Eichler: Yes, moving from studying individuals' genes, duplications, and variations between species, to genome-wide. This was right at the time when the genome project was basically hoping to finish up in the next few years.

There was a culture shift, in 1998, as follows: The whole [publicly sponsored] genome project had been done, up to that point, methodically, slowly, BAC by BAC, fosmid by fosmid, cosmid by cosmid, people assigned chromosomes and doing their regions, reporting their results at chromosome-specific workshops. And at that point, there was a shift, essentially: [Craig] Venter. Venter saying he was going to do it faster and better and sell it as a marketable product.

So [as part of this race] I was brought to NIH—first time I had ever been there. I knew that they [segmental duplications] would be difficult [to identify], and I knew they would be important. We had done some basic analysis to see how good this working draft sequence would be as opposed to finished sequence.

I remember saying that a working draft [as opposed to a careful orderly description] would mess up duplications completely, and that we wouldn't resolve them well and it would be a disaster for my research, blah blah.

But I could tell right then that it didn't matter. They were going to sequence *lots* of clones, with sequences deposited into GenBank in the next 13–15 months. MIT at that point picked up a lot of the sequencing capacity, and

Wash U [Washington University] was committed to finished, high-quality sequencing, most in ordered maps, but not all of them were.

So after that, I got a call from Eric Lander. He said, "Evan, we're going to have all these assemblies of the human genome soon, we'd sure like it if you were willing to do a genome-wide analysis [to help determine regions of duplication]. Have you ever done a genome-wide analysis?"

And I said, "No."

He said, "Well, *can* you do it, and can you do it in 4–5 weeks?" This was around 2000.

I lied. I said, "We can do this."

I knew we could do it, but I didn't know we could do it that quickly! And so we went ahead. The sequence came in. We had to come up with a pipeline to analyze duplications within the assembly. I had an *awesome* student, Jeff Bailey, who was better at computation than he was at the bench. So we sat down and drafted what we would need—how we would do it: remove the repeats, line up the sequences, genome by genome, we'd clip—there was a heuristic involved.

And then we had to execute it, but we didn't have enough computers. There was no cluster or super-computer that we had access to.

So I walked over to Hunt's office and said, "Hunt, I have an opportunity and I need some machines. And I have a guy who is actually capable of re-writing the operating system and putting Linux on all these." So he said, "OK here's $13,000, see what you can do."

Gitschier: That's not very much.

Eichler: No! But we went out and literally bought off-the-shelf from Computer City a whole bunch of machines—I think they were Dells—and we strung them up on my lab bench—there were 15 of them. I have a picture of it somewhere. And Jeff wrote some script that would distribute the load across the machines so we could parallelize the operation.

It was the middle of summer in Cleveland. And things would go for a week and crash. And the process was such that you'd have to start all over again. Two weeks in—crash.

Gitschier: Power outages?

Eichler: No—heat. The rooms weren't air conditioned *enough* to deal with the heat that was generated from 15 computers strung together, side-by-side. We had a little maelstrom of heat. So the critical component for this first cluster that we built was a Kmart fan—actually three of them—that we stuck in the back and blew the heat away from the back of the chassis of the computers, and it finally ran to the end. We ended up a little bit late, two weeks late I think, but we did our first genome-wide analysis.

And it was *really* disappointing. We realized that the first assemblies had screwed up big time—something like 20% of the genome was in these blocks of duplication, and 90% were false positives. We were bummed out because we had put all this energy into a duplication map. We had all these ideas for evolution and disease, and we realized that we didn't have it yet!

So we reported this back, and they were like—OK, we've got to fix this. So we gave all the coordinates to [David] Haussler [developer of the UCSC genome browser]. And additional assemblies went on to be more rigorous.

But to us, it wasn't what we wanted. Which were the false positives, and which were the real duplications?

So what we did—and here is where we got into a little bit of trouble—I knew Venter was doing his genome assembly a *different* way, and I knew that his assembly method would *miss* the duplications *completely*, because they couldn't actually assemble within a duplicated region. There would be "mate-pair" violations [mismatches of two shot-gun end-sequences], and they would just throw out the discrepant reads.

So we came up with this idea, which is fairly simplistic. We knew that the best part of the public [genome sequencing effort] was that they had individual haplotype BACs—150,000 base pairs in individual clones—that were good for orderly assembly. And, we knew that the best part of Craig's was this whole genome shot-gun approach. So if we could take the raw data from both projects and merge them—we'll use the depth of coverage [from Craig's shot-gun sequences] as a dipstick for duplication, and we'll take every BAC [from the public genome project]—all 36,000 of them—and we'll align all of Craig's reads against them. Wherever we see excess depth of coverage and wherever we see excess divergence will indicate a potential duplication.

So we tested it, and we had a 95% hit rate on our duplications. We could detect duplications that were big. By 2002, before any [final genome] assembly came out, we had a duplication map, and that was published [in *Science*] as a map for future studies.

And that's where I got into problems, because I was analyzing both public and private human genomes prior to either being published.

Gitschier: So you had bought into the Celera database?

Eichler: I didn't. I collaborated—I collaborated with the public and I collaborated with Venter at the same time.

Gitschier: Were there others who did that, too?

Eichler: I don't think so, because I remember getting phone calls warning me to be careful!

I was interested only in the scientific question, not the politics. And I explained that to a number of people including Francis [Collins] and Eric [Lander], and Mark Adams at Celera, with whom I had an established collaboration. I had to keep a wall of China up between the two sources. People eventually understood that I wasn't contaminating the well, on either side.

There were two things working in my favor—most of all I was naïve, and I just didn't understand a lot of things that had gone on with Congress, and things that had gone on trying to stifle one project versus the other. The second thing going for me was that I was blinded by getting the duplications sorted out. This would be the greatest thing since sliced bread in my life.

Gitschier: And it was!

Eichler: It was great, and fun. For us, that was a watershed moment. We used that information to predict regions of rapid evolutionary turnover as well as regions that we believed were disease hotspots, like the autism locus at 16p11.2. And we decided to systematically go in and look for structural variation in these regions.

Gitschier: Tell me more about the duplications.

Eichler: If you try to reconstruct the whole evolutionary history of the duplications themselves, what you notice is a couple of things. One is that humans have too many interspersed duplications compared to other mammals.

Point number two: the duplication architecture is very complex, suggesting that there has been a series of events creating almost every duplication block, of which there are about 400 in the human genome. Most of these have been created over a period of 10–15 million years, where it seems that most of the activity was around the time of the common ancestor to human and chimp and gorilla.

And then here's the kicker: if you reconstruct the entire history of these, they provide a framework. Because duplications swap material between them, they share evolutionary history. You can build a tree of relationships of the segmental duplications.

What you see is that most of the expansions have occurred on about half a dozen human chromosomes, and most of these expansions lead to these architectures, such as on Chromosomes 16 and 17, where you now have *big* blocks of duplication flanking a region and sensitizing it to microdeletion and microduplication.

So, you ask yourself, "Why these big blocks?"

Here's what I think, although we haven't definitively proven it yet. If you look at the centers of these big blocks of duplications—what we call the cores—these tend to carry rapidly evolving genes embedded within them.

So we're coming to a new and what I think is an important paradigm here: Architecture that is predisposing to microduplication and microdeletion is there at the benefit of having these newly minted genes propagating and expanding across the human genome.

There are now half a dozen gene families that have been published—some by us, some by others—which show these signatures, and most of the duplication architecture seems to be almost a genetic hitchhiker—these cores that have landed in new areas, picked up flanking material and duplicated again to other sites.

I would have to think that solving this riddle of what the function of those genes are would be tremendous. The genes that are in there tend to mark the oldest and the deepest part of the duplication block. The block itself is a mosaic of different pieces. There are bits of pieces of genes, some are transcribed, some are not, most are neutrally evolving. But the cores carry genes that show strong signatures of positive selection. These are genes that are smack in the middle of hundreds of kb [kilobases] of complex duplication territory, where Affymetrix, Agilent, and the SNP people have feared to tread! This is the "un-HapMap-able" region of the human genome. And if there were any association with any disease, people would have missed it because there is no type of genotyping technology to actually assay.

And here's the rub—these genes are not only embedded in complex duplications, they are even copy number variant between humans. We have a gene family we call "Morpheus", which we published in 2001. Some people have 20 copies, some have 16 copies.

So, what are the functions of these genes? Tough question! This is the geneticist's worst nightmare: Mice don't have them, so you can't knock them out. There are multiple copies in humans, so would be tough to genotype. They are too far away from any flanking tagged SNP to find any type of association. So there are a whole series of black holes.

Gitschier: What has been the thing about your research that had you the most jazzed?

Eichler: I'm still jazzed!

Four or five years ago I made a conscious decision to go back to my roots—to go back to human disease. Up to that point, I was focusing on human duplications strictly from the perspective of structural variation in humans and variation among primates. So taking what I learnt from David Nelson and Jim Lupski, I've come full circle, because we are now studying children with disease, and we're finding what we think are causes—at least associations—now. But we're doing it from the perspective of looking at the genomic architecture, as opposed to linkage or association.

Both of those make me feel good—I love the evolutionary history of the human genome, and I'm completely unapologetic that I am anthropocentric—that if I were doing the same thing in *Drosophilia*, it would not interest me. But in humans—I care how we tick! That we can walk around with all of this *stuff*. It's almost liberating—the fact that there is no perfect genome—that all of us are made up of deletions and structural changes and copy number variations.

It's amazing that any of us are "normal". And maybe none of us really are—and that's the beauty of it!

The Exception That Proves the Rule

An Interview with Jenny Graves

In addition to the eutherian mammals, such as dogs and elephants, with which we are most familial, the class Mammalia has two additional branches: the marsupials, including kangaroos and possums, whose gestation largely occurs in an external pouch, and the monotremes, of which there are only two extant species (the platypus and the spiny anteater), who also lactate but lay eggs. Genetic analysis of the marsupials and monotremes has proven extremely important in generating ideas about the evolution of mammals, particularly in the areas of the mammalian sex chromosomes. Even more captivating is the recently sequenced platypus genome, which harbors genetic fossil records of both reptilian and mammalian origins. This interview delves into some of these thought-provoking discoveries.

Interviewed November 12, 2007

Published June 27, 2008

CLOSE TO 20 YEARS AGO, I WAS CONTACTED by an Australian woman who was planning to map the locations of genes that are X-linked in humans in some odd Australian critters, the monotremes. These animals comprise a distantly related branch of mammals that have hair and lactate, but additionally lay eggs. She wanted a probe from our lab, and, in exchange, little vials of DNA from spiny echidna and platypus appeared in the mail. Our lab became enamoured of these singular animals, and we followed their scientific story with great interest. The lady was Jenny Graves, and it has taken me this long to finally meet her.

Jenny is a native Australian, and one of her first dips into research as an undergraduate honours student involved asking whether kangaroos employ X-inactivation [they do]. Although she deviated from this interest as a PhD student in Berkeley and fully expected to follow a different research path, she found herself back in Australia, post-PhD, where a colleague suggested she look at gene mapping in marsupials. Though her curiosity about X-inactivation has been the driving force for much of her work, this serendipitous suggestion proved pivotal. Jenny's work on monotremes and marsupials has led to powerful insights into the evolution and function of the X and Y chromosomes in mammals.

Jenny has had a long and joyful career, overcoming a near fatal illness and a coincident collapse in funding to resurrect her research, and eventually being awarded the L'Oreal Prize for women in science. She is now head of Comparative Genomics at the Australian National University in Canberra and Director of the Australian Research Council Centre of Excellence for Kangaroo Genomics. I was able to interview her during a visit to her daughter, who lives in the San Francisco Bay area. It was a sunny November afternoon in an outdoor café on the island of Alameda, and over a plate of mussels, some wine...

Gitschier: So much of your work has revolved around your interest in X chromosome inactivation. You've worked on methylation and transcription of X-linked genes in mammalian cells. What is the story in marsupials?

Graves: It turns out that in marsupials, DNA methylation doesn't seem to be important! Marsupials are always a shock. Just when you think you know something, you do the same experiments in marsupials and you get the opposite result.

Now we have a genome sequence [from opossum and kangaroo], so we know what genes are on the X and we can just go down the chromosome and ask—are *you* active, are *you* active?

So this is sort of the end—40 years later—of a very long dream of trying to find out how X-inactivation evolved.

Gitschier: It's probably still not the end!

Graves: I was just thinking I shouldn't have said "end." But X-inactivation is what excites my curiosity more than anything.

Gitschier: And that's how you started out, too.

Graves: Yes. After that, I went to Berkeley to do my PhD and I had no intention of working on marsupials ever again.

When I came back to Australia, my friend Des Cooper said, "Why don't you map some genes in kangaroos?" I wasn't interested in kangaroos at all. But just to be nice, I did that, and it turned out to be terribly interesting, because the first three genes I looked at were on the X chromosome, like we thought, which was the first indication that the X chromosomes were monophyletic—that they share a common ancestor.

But the next few we looked at weren't! Which was a big shock. It turns out that all the genes on the short arm of the human X are on one place on an autosome in kangaroo. What on earth does that mean? That immediately told us that the human X had an ancient bit and a recent bit.

And that was interesting because it solved a lot of problems. The human X is really strange because the top bit doesn't act like an X at all. It's full of genes that aren't inactivated, and that's because they haven't been on the X for very long.

So that told me something that's been with me my whole life long, which is that sometimes, when you ask a functional question, you get an evolutionary answer. And this has become a guiding principle. Sometimes things are the way they are not because they work better, but because that's the way they evolved. And you can guess function by asking a gene where it has been in the last 100 million years.

Gitschier: It must be very thrilling to take a look at something that has been around for a hundred million years and make that observation for the first time.

Graves: Absolutely. I started out in molecular genetics and was greatly in awe of the evolutionary biologists who seemed to have their heads in the clouds. I didn't think we even belonged on the same planet! But it never dawned on me how relevant evolutionary thinking was. I didn't realize how all of the answers come from evolution. It's been a real thrill to plug my work into a much bigger framework of how genomes evolved.

Gitschier: What do you make of the fact that monotremes exist only in Australia?

Graves: They probably didn't. There is a record of one tooth—one monotreme tooth in South America. It is the most amazing story—I just don't know if I believe it or not. To begin with, monotremes just don't have teeth!

But the ancient monotremes *did* have teeth and then they lost their teeth and now have grinding pads instead. But they do have a hatching tooth, so it is still possible to compare with the fossils.

The story I heard was that somebody sat on a tooth and said, "Ouch, what is this? A monotreme tooth!" Well, how on earth would even a paleontologist know a monotreme tooth? Apparently they are very, very distinctive. And it is pretty much accepted that there were monotremes in South America. Whether they evolved in South America or Australia or what is now Antarctica we don't know. But obviously they spread into Australia and New Guinea, which also has a number of species of *Echidna*. But the platypus is unique to Australia. There were none in New Zealand. But there is this one tooth in Argentina!

Gitschier: Now for a silly question: What's it like to hug a koala?

Graves: Very uncomfortable. I still have the scars from Bonnie the koala from some photo shoot. I've hugged many koalas, but Bonnie put her arm around me and went *ngghh*. They are not nearly as cuddly as they look.

Gitschier: Are there any other marsupials that you've had an intimate relationship with?

Graves: I don't do a lot of animal work. I've certainly held a lot of young kangaroos. The little ones are very docile and incredibly cute. But most marsupials are not very good pet material.

Gitschier: So none of them have been domesticated.

Graves: No, but there is a lot of interest in using marsupials more, particularly kangaroos in the meat trade, because Australia is drying out and sheep and cattle are terribly damaging to the fragile soil.

Gitschier: How are the monotremes doing—are they endangered?

Graves: We call them "vulnerable", but that's because the platypus is largely aquatic and it is terribly sensitive to the quality of the water. But the farm runoff has been fixed and, in areas around Melbourne where I live, the platypus is coming back, even in suburban Melbourne.

Gitschier: Can you see them in the wild? I had to look very carefully even in the aquarium at the Sydney zoo. They are much smaller than I had realized.

Graves: It is difficult. I've been on many muddy riverbanks on very cold mornings. There are some places that I can almost always show them to visitors. It's a big thrill even if you don't see too much: a hump in the water and two little eyes looking at you.

It's very difficult to breed the platypus in captivity. One bred in 1934 and then nobody could get any to do it again for another 70 years until they got a new platypusary, with lots of room with long tunnels. Seems to be this one female platypus that has given birth three times. She lays two eggs but only one hatchling survives usually. And she stays with them for 3 months. She makes milk—very complex milk, but she has no teats—it's just exuded from the skin on the abdomen. The young just lie on the skin on the abdomen and lick the milk from her fur.

Lactation is very ancient, but the mammary gland is just a glorified sweat gland. And the mother just lies there for months without getting in food, so she has to be in very good shape.

It was really easy to convince the NIH to sequence the platypus genome because it is so unique—and it is the link between mammals and reptiles. There are so many things about the platypus that are reptilian. It is a mammal—it has fur and it makes milk—but it also lays eggs, and it has a very different structure of the embryo—much more like a reptile.

And to our amazement, the sex chromosomes are more like birds'. When we looked at the sex chromosomes, we found it has ten! Five X and five Y chromosomes. Now we know what genes are on them—they have no homology to human X chromosome, but rather have homology to the bird Z chromosome!

Gitschier: And the echidna?

Graves: It also has multiple sex chromosomes that are similar to bird.

Gitschier: So you mean, to be male you have to have five X's and five Y's. And they are completely different from each other?

Graves: Absolutely!

Gitschier: How do they do that?

Graves: XY pairs have a little bit of homology. A little bit of X1 has homology to Y1 and then Y1 at the other end has a little bit of homology to X2 and then the other end of X2 is homologous to Y2. So these pseudoautosomal regions pair, and at meiosis you can actually see ten chromosomes in a chain—XYXYXY, etc. How they segregate is a mystery—we've never actually caught them at anaphase, but we think all the X's must go one way and all the Y's the other, because we've never actually seen a sperm or spermatocyte with both X's and Y's in them.

But we're not so surprised because there are multiple chromosomes in some plants like the evening primrose, and in some spiders. It seems to have happened when two different chromosomes swap bits and must pair in a chain of four, and then one of these swaps bits, and so on. Functionally quite crazy, but once it happens, it is stuck, and must make the best of it.

It really amused me to be told once that our *Nature* paper on platypus sex chromosomes was featured on the "Discovery" Web site. And I said, "Oh, that's wonderful," and they said, "Well maybe you don't know that the Discovery Web site is creationist, and your paper is put on there as an example of intelligent design!"

I said, "That's the dumbest thing I've ever heard!" And that was the inspiration for my "dumb design" Web site, which I'm setting up now with my L'Oreal prize money, as examples of how evolution can explain things very simply that seem to make no functional sense at all. So that's going to be my first example, of something that happened once, accidentally, that now can't unhappen, but how systems work around these accidents to make the best of a bad job.

It truly distresses me to see kids being brought up to believe in utter nonsense [creationism/intelligent design].

Gitschier: Is that true in Australia, too?

Graves: Not as true in Australia as in the US. But there is a lot of pressure to accept the teaching of utter nonsense in school. I think we are raising a credulous generation who will believe anything as long as they read it in *Reader's Digest*. It's so dangerous to encourage people to believe what they are told rather than what they observe.

Over many years I've been distressed to find that our students come to us without the ability to observe anything! Our students will sit in front of microscopes and draw things that aren't there! And I'll say, "Well, that's very nice to draw chromosomes with a spindle attached. Do you see a spindle attached?"

"Oh yes!"

Of course, you can't *see* spindles! So I look down the microscope and say, "I don't see a spindle." And they say, "But I know it is there, because my textbook has a spindle."

As long as you are drawing things that aren't there because somebody tells you they are there, you are in deep, deep trouble.

And I think it is much deeper than just believing or not believing in evolution, you've got to change education to encourage children to go looking for themselves. And start thinking to themselves: well—who's right?—what I see or what someone tells me?

So I'm becoming very interested in education, particularly of young children, which is where I think the rot sets in. Science is not taught well even at high school level, and at primary school level it is taught by people who are generally scared of science! Anybody who has anything to do with kids this age knows they are incredibly observant and incredibly clever at working out how what they observe relates to other things. Somehow that just gets lost. I'd love to see more attention on encouraging young kids to make their own hypotheses—crazy though they may be.

Gitschier: Let's turn to the testes-determining factor and the race to clone the gene. I'd love to hear your recall of that period.

Graves: I'll tell you the story as it happened because it is a good yarn.

I had no pretensions of working on sex-determination at all. But of course I was interested, and I was watching this *war* going on between David Page's group in Boston and Peter Goodfellow's group in London regarding finding the testes-determining factor.

When David Page's paper on ZFY [the putative testes-determining factor] came out in *Cell* in 1987 I thought, "Wow, this is gorgeous, beautifully done." I didn't think we would have anything to contribute. But David then called me up that same night. I didn't know him at all. He said they would really like to show this gene is on the Y chromosome in marsupials. So he sent us the probe.

Oddly enough, Peter Goodfellow sent us an independent probe of the same gene. He just sent the probe, he didn't even ask in advance!

My student Andrew Sinclair—it was the last week of his PhD work—had looked at other genes on the short arm of the X chromosome in humans and had shown that they are autosomal in marsupials.

The gene David sent us [which came from the Y chromosome] also had a friend [a homolog] in that short arm of the human X chromosome. We thought: that's interesting, wonder where ZFY will be in marsupials.

Well, Andrew called me up late one night and said, hope you're sitting down because ZFY is not on the Y chromosome: it is on Chromosome 5—which is a very funny place to keep your sex-determining gene.

So I said, "Don't be silly, look at more cells." This is the old days where we used radioactive in situ hybridization and had to count silver grains over hundreds of cells. The next morning he had absolutely incontrovertible evidence that it was in the same patch of autosomal genes that we had shown should have been on the X, but weren't.

So I told David, and of course he didn't believe it, because that meant that ZFY was not the testes-determining factor. He thought there was just something very strange about marsupials.

In the meantime, we got the same result with the probe Peter had sent us, and he was *very* keen to publish [it became a cover article in *Nature*]. Andrew later went to Peter Goodfellow's lab, which had been organized before all this had happened, and it was he who cloned the SRY gene, which was the true testis-determining gene.

I was right in the middle of the war zone. For an innocent little Australian, that was quite a wake-up call! But I've become good friends with both of them.

Gitschier: About this period when you lost a lot of funding and had your illness— I'd like to talk to you about that because it may be inspirational to others of us when we go through troughs.

Graves: It was strange. My lab was sailing along. We had our cover story in *Nature*, and another paper in *Nature*. I didn't have a care in the world. Then, I failed to renew two grants. Maybe I was just too cocky.

When a week later I collapsed from a brain bleed, the rumor went around that "Jenny's died from a broken heart!" Everything seemed to be conspiring to say, "That's the end of you!"

But I was fortunate. I had a great neurosurgeon and a wonderful collaboration with Art Riggs in Los Angeles on the platypus, and he funded a technician in my lab. I called him from my hospital bed, telling him I was in dire trouble and asking whether he could support her for another year. That absolutely saved me—to keep that expertise in the lab! And the university gave a scholarship to another technician in my lab, so I could continue. And I did have two loyal graduate students.

This was 1992–1993. Arterio–ventricular malformation, a congenital thing, which is not all that uncommon, but it was in a very bad place—the fourth ventricle. The neurosurgeons told me it would take me 18 months to get back on my feet, which was true. I couldn't see, I couldn't walk, and I was being sick all the time.

I had a lot of time to think, and I could type! I had a proofreader. I wrote five grants and I got the lot. From three people in the lab, I had 18 the next year.

Gitschier: During that period, though, did you consider doing something completely different?

Graves: People said, "Now is the chance to really think about your life and what you want to do." I did—and I thought, "Yes, I want more of the same!"

I did think about other things I could do, particularly if my vision was to be impaired, but I very quickly decided I love what I'm doing and couldn't wait to get back to the lab, and I signed two contracts for books, one of which I'm still working on—the molecular biology of sex chromosomes. I had so much fun. I found I knew so much I could just write it without referencing anything, and then when I recovered, I started plugging in real data and references.

Gitschier: So you may not retire.

Graves: I will retire from writing grants, I swear. I have enough grant money till 2010. But I'll be 69 by that time and I'm not sure I want to keep running a wet lab. It would be a good time to return to Melbourne, where I live, and I'll continue writing. I think I'll be as happy as a lark and co-supervise students at the Uni [University of Melbourne]. I want to learn to be a bioinformatics person. The world is full of data.

Gitschier: When did it occur to you that studying marsupials and monotremes would help to solve such important questions about evolution?

Graves: I'm ashamed to admit it didn't dawn on me right away. I found it fascinating to map genes in marsupials, but it still didn't get to me that this was *important*.

The beginning of that realization was when Steve O'Brien came to Australia in 1984. He saw right away that animals distantly related to human would be extremely powerful. So it was really he who convinced me to believe in what I was doing. And once I had started on that track, it was easy to find more and more things that one could look at. Anything you like can be looked at through the spectrum of evolution, and if you have systems that are so divergent, comparisons are extremely powerful. And of course, that was before sequencing, we were just looking at arrangements of genes, simple mapping. But already it was obvious that the out-group gave you real power to tell you how the genome was rearranged.

Gitschier: You mean monotremes.

Graves: Monotremes are an out-group to marsupials and placental mammals, marsupials are an out-group to placental mammals. Chicken is the out-group to all the mammals.

All of a sudden I realized I was sitting on a gold mine. And I felt such a sense of responsibility. Here is a very Australian gold mine, and I've got to get out and beat the hedges and tell people what we've got in Australia. Because Australians, curiously, don't realize what we could be doing with our own native flora and fauna. It is not a well-funded field.

Gitschier: So, in a sense, moving back to Australia was the best career move you could have ever made.

Graves: It was! I didn't think of it that way. I never expected I could be doing something unique and important so far away from the action.

I always tell young Australians, knowing there are such huge resources in the north [Northern Hemisphere] that we are competing with, if you can find something unique, then you know what you are producing is unique. Then you just have to worry whether it is unique and *boring* or unique and exciting. And of course, you don't know, sometimes you have to do some work to find out. You can pretty soon see whether you are getting general principles out of it.

We were very lucky. The whole ZFY business really showed me that we were on the right track.

I used to give talks on kangaroo genomes in the US and people would laugh! If I showed a koala picture they would say "Ahhhhh", and if I'd put up a platypus they'd fall off their chairs laughing. I thought, "At least they're paying attention."

After the SRY story, nobody was laughing at kangaroos any more.

The Eureka Moment

An Interview with Sir Alec Jeffreys

It is now recognized that any two human beings differ by about 1 in 1000 positions in their DNA, and it is by virtue of these differences that we can uniquely identify each person or each pair of identical twins. Regions of the human genome with short repetitive sequences are hotspots for genomic variation because the repeats can expand and contract easily. The unexpected discovery of highly variable regions and their early use in identity testing is discussed in this interview. In addition to so-called DNA "fingerprints," the interview explores the serendipitous finding of intervening sequences, also known as "introns," that disrupt the coding information contained within genes.

Interviewed June 18, 2009

Published December 11, 2009

IN 1984, WHILE TRACKING THE VEINS OF GLOBIN GENE EVOLUTION and panning the human genome for hypervariable linkage markers, Sir Alec Jeffreys accidentally struck gold—he discovered a way to identify any human being by a DNA "fingerprint". To use Jeffreys' words, he has been "branded" by DNA fingerprinting, but he delights in its application and the hook it provides for public curiosity about science. Like Jeffreys himself, I wanted to dig below the surface of this discovery as well as that of another genetic nugget—the intervening sequence—found as a post-doctoral fellow seven years earlier.

On the heels of my interview with Adrian Bird (published in the October issue of *PLoS Genetics*), I made my way to Jeffreys through another branch of the British Rail system. When I arrived at his building on the leafy Leicester campus about 45 minutes early for our appointment, his assistant suggested I get a cup of coffee while Jeffreys finished his experiment. I certainly wouldn't have needed one. Jeffreys is an animated speaker, with a resonant voice and a rapid delivery of succinct clauses strung together in run-on sentences. His story could have cut through anyone's jet lag.

Gitschier: I didn't realize that you still work in the laboratory.

Jeffreys: I certainly do!

Gitschier: Tell me about the experiment you were just doing.

Jeffreys: Right, well, we won't go into the gory details. Copy number variation [CNV] in the human genome is a real hot topic at the moment.

Gitschier: The kind of variations people are looking for in association with autism and psychiatric diseases.

Jeffreys: That's exactly right. It's a common phenomenon, and we've actually known that for decades. What we're doing is going back to some of the absolutely classic examples of CNV. These are in my favorite gene family—the globin genes—and that's where I cut my scientific teeth.

Gitschier: We're going to be coming back to that!

Jeffreys: Right. So, what I've done in my scientific career is this gigantic circle, starting off in globin genes, going all around the place in forensics, and returning back to my first love. The experiment I'm doing at the moment is looking at de novo copy number variation in the fetal γ-globin genes at the single molecule level in both somatic and germline DNA.

All of this comes out of my work on recombination hotspots. And the general feeling was that recombination hotspots function at meiosis—they drive allelic recombination, and they may well drive ectopic recombination.

Gitschier: Define "ectopic."

Jeffreys: The term ectopic originally came from yeast and it applied there to a situation in which you have a sequence repeated, say here and there, so that they can undergo unequal crossover and cause duplication and deletion. "Ectopic" recombination means it's "out of place."

As of yesterday, I found there is copy number instability not just in the germline, but in somatic DNA. That largely rules out meiosis and meiotic recombination hotspots. Even in the germline it is quite clear that the substantial proportion, possibly the great majority of rearrangements, are again pre-meiotic, arising during germ cell development. We're trying to drill down below the applied genetics [looking for variation associated with disease], to some of the fundamental mechanisms, to understand the dynamics of rearrangements in the human genome.

So, if you want to put a simple summary on what this lab is about, it's about human DNA diversity and the processes that generate it.

Gitschier: OK! Now let's get to the first question on my list, which indeed is about globin. It's about the period of your post-doc in Amsterdam. Why did you go there and why work with Flavell?

Jeffreys: OK. I did my D. Phil. at Oxford University on human somatic cell genetics. Then went to a Biochemical Society meeting and chatted with a chap named Piet Borst, a very senior scientist, who at the end said, if you are interested in doing a post-doc with me, just let me know.

And I thought, that's great, 'cause I wanted to get out of Oxford, and Holland I really fancied because the language wasn't going to be a problem; everyone speaks English. So, I got myself an EMBO fellowship to work with Piet on yeast tRNA genes.

In 1975, the door was clearly opening on molecular genetics; before that, it wasn't worth talking about.

Gitschier: Expand on that statement.

Jeffreys: I remember very clearly. There was a colleague of mine at Oxford called David Finnegan and we're waiting in the lunch queue and he wanted to go off to the States, and I said what's the project, and he said the idea is to take *Drosophila* DNA and to try to stick bits of that into lambda phage.

Gitschier: With David Hogness?

Jeffreys: Right. And the penny dropped then, that this was going to be the way forward.

I get to Amsterdam, and Piet said, you can work on this if you like, but you might also like to have a chat with this guy Dick Flavell, he's got a collaboration with Charlie Weissmann in Zurich, on trying to isolate a mammalian gene. And I thought—whoa! That sounds really exciting. The idea of the project was to get to a single-copy gene. No one had ever done that in a mammalian system. The only one we could possibly do, we felt, was either rabbit α- or rabbit β-globin, because the mRNA had been purified. The gene isolation would be by physical purification.

Gitschier: No cloning?

Jeffreys: Well, cloning came in right at the end. It simply wasn't around at the time. It was by hybridization enrichment with prodigious quantities of DNA [from rabbit liver]. The experiment was to cut it up with EcoRI restriction enzyme. Remember, this is back in the days when you couldn't just buy enzymes off the shelf, you had to *make* them.

Then denature the DNA and hybridize it to globin mRNA. This was a two-pronged attack. In Amsterdam we were going to use the mRNA to pull out the complementary strand, heavily enriched, and in Zurich, Charlie Weissmann had managed to make a cDNA so he could pull out the other strand, and the idea was to purify our complementary strands and then meet somewhere in the middle to hybridize the two stands back together. Then, because this was an EcoRI fragment, we could then pop it into a vector that we hoped someone was about to develop.

Gitschier: How were you selecting the mRNA?

Jeffreys: We were selecting by attaching mercury to the RNA and then capturing it on a thiol column.

Gitschier: That's a dangerous experiment.

Jeffreys: Oh, the whole thing was horrendous. We were using radioactive mercury.

Gitschier: But hold on. Since there was no reason to suspect that there were intervening sequences, what is the point of going after the gene?

Jeffreys: Nobody had ever seen a mammalian gene. No one had any idea of what it would look like.

Gitschier: So, the idea was to get something bigger than the mRNA itself.

Jeffreys: Yes, that's right. To look at the flanking regions. Basic academic curiosity.

During that experiment, we had to develop methods for monitoring purification, and the only way we could see to do that was to use Ed Southern's blotting technique, which at that point was only a year or two old.

So, as we purified the DNA we could monitor the fractions just by running them out on an agarose gel, doing the Southern blot and then hybridizing with an appropriate complementary probe. And that not only worked, but we could actually *see* the fragment of DNA we were trying to purify in the starting EcoRI digest of genomic DNA.

Gitschier: Hadn't he shown that before?

Jeffreys: No, Ed was desperately trying to get this going. I know Ed extremely well, and there was a bit of discomfort on my part thinking that we had trampled on his patch. On the other hand, that is what we needed to do.

Having got the ability to detect down to the single gene level, we thought we should see if we could make a restriction map around the gene, which is what we did.

Gitschier: Were there EcoRI sites in the cDNA?

Jeffreys: No. The cDNA had been cloned by Tom Maniatis, and we pretty quickly moved over to using his rabbit β-globin cDNA that he very generously provided to act as a probe for monitoring. We just wanted to check that everything was OK. And we built up a restriction map around it [on genomic DNA via Southern blotting].

We then discovered that there was an EcoRI site right smack bang in the middle of the gene! [That meant] our enrichment experiment was a total disaster, because we would have purified one end of the gene in Amsterdam, and in Zurich, they would have purified the other end of the gene, and to put them together, there would be nothing. The flop of the millennium that was!

But, the question then was, what the hell is the EcoRI site doing in the middle of the gene? And then we started to do more and more fine-mapping and it was clear there was a huge gap in the gene.

I remember sitting down with my Dutch technician, saying we've got the restriction mapping data, let's try putting all this together. And I knew it was just nuts, but I thought we could solve it if we just put an extra dollop of DNA inside the gene. All of this was done without reference to Phil Sharp and Rich Roberts's work with adeno [which was happening at the same time]. I knew instinctively that this was something pretty exciting. And then, Dick was over for, I think, a Cold Spring Harbor Meeting, and everything started falling together. About the same time, Phil Leder managed to clone in lambda the mouse β-globin gene and showed by electron microscope analysis that there was additional sequence inside the gene. But the trouble there was it had been cloned in *E. coli* and perhaps it was an insertion sequence. And then [there was] Chambon's ovalbumin gene story.

Looking back on it, basically in 1977, introns were going to be discovered. Full stop. The technology had arrived to the point where the discovery was inevitable. I think all of us in the field were grateful that we just happened to be at the right place at the right time.

When it was time to leave Amsterdam, one possibility was to do a post-doc with Ed Southern up in Edinburgh. He's a great guy and the stuff he was doing was fantastic. He's one of my heroes. We are actually quite similar. We like fiddling around with things. He gave this wonderful quote a few years ago that he misses the days when he could get at the data before the computer did.

But at the same time, I thought I'd like to try running my own lab, and out of the blue came a phone call from this guy called Bob Pritchard who founded this Department [of Genetics] in the early '60s. He said, "Would you be interesting in coming for an interview?" I said, "Where is it?" He said, "Leicester." And I said, "That will be fine."

I put the phone down and I said, "Where the hell is Leicester?" All these Dutch people were running around trying to find a map of Europe.

Gitschier: Pre-internet.

Jeffreys: Pre-everything! These were the days if you wanted a sequence you had to get out a typewriter and type it in.

So, I visited Leicester and I immediately fell in love with the department. I came as a temporary lecturer, and I'm still here 32 years later, so it says something about the environment. I love it here.

So, the question then was, what was I going to do? It was clear that carrying on with the intron work was not going to be viable. Suddenly everybody was moving into the field—evolution of introns, mechanisms of splicing, etc. I thought, take your education in human genetics and your new-fangled molecular biology and stick them together. If you can pick up specific bits of human DNA, then you should be able to scan for variation. Variation that affects a restriction enzyme site will manifest as what is now called, I think very uglily, an RFLP [restriction fragment length polymorphism].

So, that was our first quest. By early 1978 we had picked up our first RFLP, a rare variant in a single individual. Again, these were in the globin gene clusters, because again, these were the only genes for which probes existed at that time. Really excited, but we got pipped to the post because Kan and Dozy published their RFLP and the association with sickle cell disease.

Gitschier: I think they just bumped into that discovery.

Jeffreys: What we had done was to do a fairly systematic survey for RFLPs in the β-gene cluster.

Gitschier: What made you think that there would be variation in restriction sites among people?

Jeffreys: I can't remember. It seemed fairly obvious at the time. I knew enough human genetics to know that there must be a significant amount of variation in DNA sequence. I'd been brought up in the days of serology and biochemical genetics, enzyme polymorphisms, and we knew that that was sampling only a tiny proportion of all diversity in the genome. So, if there is diversity, then it will be agnostic with respect to restriction sites, so if you luck out, you'll find a polymorphism that hits a restriction site and that makes it assayable.

Having come up with these RFLPs, we then got fed up with them, cause everyone was doing it. So, we then started thinking that surely in this enormous human genome, there must be bits of DNA that are more variable than these RFLPs, and we thought intuitively that the right place to look was tandem repeat DNA. I've been brought up in the school of satellite DNAs, which was the only class of DNA you could purify going back to the old cesium chloride density gradient days. The satellite DNAs incidentally show a lot of variability in copy number.

I felt intuitively that if you had local tandem repeat sequences on a smaller scale in the genome, they'd have potential variation as well. The hypothesis was that there may be bits of DNA with repeats, maybe 10 or 20 bases long repeated 10 or 20 times, so we started all kinds of crazy experiments trying to physically purify these bits of DNA from the human genome.

Then in 1980 Arlene Wyman and Ray White described the first hypervariable locus, so I thought, WOW they do exist! But their interpretation was one of transposition. Why? Because they came from a transposable element background. So, quite reasonably, they were thinking, OK it's hypervariable because we've got a transposable element that is moving in and moving out, taking DNA with it and creating this length variation. But, I read their interpretation of transposition and I just felt not so sure about that. So, we then started redoubling our efforts and still getting nowhere at all.

Then Graeme Bell described the sequence of the human insulin gene and right next door to it was a minisatellite—a highly variable tandem repeat region. And then Doug Higgs in the α-globin region.

Gitschier: What approach were you using to try to find these variable minisatellites?

Jeffreys: It was primarily physical enrichments. These sequences might have unusually fast reannealing kinetics, so you could do a COT approach. Or, since these sequences might be quite long but consisted of repeats over and over again they would tend to be resistant to restriction enzymes, so, if you took a load of common cutting restriction enzymes, you would leave these things intact.

We were still getting nowhere. But meanwhile [in a separate project], we were doing some globin gene family evolution work. We thought, OK there is a missing gene in the story, and that is myoglobin. Could we get the myoglobin gene out and see how it fitted in to the hemoglobin gene family as a very diverged member of that family?

So, this is really the start of the DNA fingerprint story, because we got the human myoblobin gene and found a minisatellite inside the intron.

Gitschier: How did you find that?

Jeffreys: By sequencing. It wasn't variable between people, but I realized I had seen this sequence somewhere else. So, I went back and looked at the α-globin and the insulin minisatellites, and you could see this sort of vague suggestion that there might be some sort of shared sequence in there. So, we then took that myoglobin minisatellite and hybridized it to a human lambda library and lo and behold a number of clones lit up. We then started systematically isolating those clones, showed that they contained minisatellites and some of them were pretty variable loci.

Gitschier: So, you were checking this on a Southern blot?

Jeffreys: Southern blot and characterizing by sequencing. And, as we were building up the repeat sequences from the clones coming out of the library, the shared sequence motif, the minisatellite core, became more and more obvious. It was a short sequence, about 15 bases long, embedded within the repeats of the minisatellites. It was almost as if this was some kind of sequence driving this tandem repetition. But, more important, it could give you a much more effective generic way of getting minisatellites out of the genome, because rather than using this crummy myoglobin probe, you take a probe that consists of just this core sequence repeated over and over again.

So, we took that and hybridized it to a Southern blot, which happened to have [DNA from] the lab technician and her mom and dad. We got this fuzzy splodgy mess, but the DNA fingerprint was absolutely obvious. We got a pattern like a fuzzy bar code. These patterns were individual specific, and seemed to be inherited within the family. That was a real eureka moment, because we were suddenly onto something completely new, which was DNA-based identification.

Recall, the driver for this experiment was medical genetics. You needed these improved markers for facilitating construction of linkage maps of the human genome and helping in linkage analysis of inherited disease. This thing would have been useful were it just a single location in the genome, but the fact that there were multiple copies of the repeat sequence in the genome gave it a new meaning, in terms of DNA identification.

When I talked about it in a Department seminar, and then speculated about what we could use this for, like catching rapists from semen—about a third of the audience fell over laughing. It sounds bizarre now because it's so blindingly obvious that you can use DNA for this, but believe me, back in the '80s it was simply not there. The only reason I came up with the idea of DNA-based identification was that it just hit you in the face!

So, within the first day, we saw identification, we could foresee forensic analysis if DNA survived in forensic specimens, zygosity testing in twins, paternity testing, and immigration disputes. Just like drawing up a shopping list—if we could get this technology improved, what it could be applied to.

Gitschier: I clearly remember that *Nature* paper [1985] involving the immigration dispute that you helped to settle, the case where a boy was threatened with deportation because the immigration authorities alleged he wasn't the biological son.

Jeffreys: That was the first DNA case tackled anywhere in the world, and it is still my favorite case because I was there at the tribunal where they dropped the case against the boy, when the mother was told—and just the look in that mother's eyes! She had been fighting the case for two years.

That was my golden moment. Without DNA, he could have been deported.

Gitschier: That set of events must have built up momentum for you and your lab.

Jeffreys: Oh, it did. We hadn't realized how many thousands of *other* people were trapped in these disputes! So, the next thing was a complete avalanche of letters and phone calls; people were turning up at my home!

Gitschier: What did you do?

Jeffreys: Well, I nearly had a nervous breakdown, but I kept going. It was an insane two years, 1985–1987, before the thing went commercial. We were the only lab providing any testing at all.

Gitschier: And then there is the local double rape/murder case in a village near Leicester. I read "The Blooding."

Jeffreys: It's a good book. It's accurate.

Gitschier: Ah, you've answered my question. It depicted you as this chain-smoking guy in a black jumper [sweater]. What did you think about that?

Jeffreys: Well, let me tell you a little story.

The author was Joseph Wambaugh, an ex-LA cop who happened to read about the story in his dentist's office in *Hippocrates* magazine. He thought this is brilliant, and he took the plane straight over here and interviewed all sorts of people, myself included. My secretary had written "interview with Rambo" on my calendar. No idea who he was. And I was very cautious.

He arrived, and we did not get along terribly well, talking at cross-purposes. He wanted to dig as deep as he possibly could—that was his job as an author—and my instinct was to keep stum.

It was an extraordinary case. We were approached in 1986 by the police. They said, we've got these terrible double rape/murder cases, we have a prime suspect who has confessed to the second murder. We've heard about this DNA fingerprinting and could you use this technology, not to confirm his guilt with respect to the second murder, we know that, but to have a look at the first murder and see if we can tie him in.

So, I said we'll do this, but I explained at the outset that we wouldn't be using DNA fingerprinting, but we'd use this derived technology DNA profiling, which we thought would be much more appropriate. And we said, "Don't hold your breath. No one has ever attempted this before."

Gitschier: Tell me about profiling, what it means, and why you used it instead.

Jeffreys: We knew that DNA fingerprints were too insensitive for forensic case-work. So, we simply took out the minisatellite core probe, we went back into our libraries of DNA and cloned out the most variable single locus probes, each of which gave a simple but highly variable two-band pattern. We knew that was the way forward.

Gitschier: OK, back to the case.

Jeffreys: The forensic samples arrived, and I have to say that was a chilling moment. An ordinary academic and suddenly you've got murder samples in front of you. I remember my blood literally running cold at that point.

We put the first probe on, and the prime suspect wasn't a match [with the semen sample from the second murder]! Suddenly we were into the world of exclusion, and how many probes do you need for that? One. The result was so wacky, so totally out of keeping from what the police were expecting to see. We thought better do another one [probe]. The results were totally astonishing, totally overturned what the police had got fixed in their minds about the guilt of this prime suspect. He was released.

The police said, OK we now believe all this DNA testing, let's go and pan the entire local community and see if we can flush out the true murderer. That was all done by Home Office forensic scientists, who at that point had our DNA fingerprinting in place. But of those 5000 samples, only 500 were DNA fingerprinted. The others were all excluded by [biochemical] testing.

The upshot of that was that the true perpetrator was flushed out, and the rest is history.

Gitschier: Have you been in any other books?

Jeffreys: I've certainly turned up in all sorts of science-y books. DNA fingerprinting is now part of the curriculum for kids age 14–15 in the UK.

So, I've achieved that sort of rare status of science reaching out to the public and being understood by school kids. And literally every 2 or 3 days I get an e-mail, mainly from the States, from school kids saying, "I've got to do a project on a famous scientist, so I've chosen you," and I love that. I always respond.

It's great because if you think you are doing even the tiniest bit to switch people on to science, and this DNA stuff is great—OJ Simpson, the Romanovs, it's got everybody. If you can't hook people into science with that story, give up.

Taken to School

An Interview with the Honorable Judge John E. Jones III

The United States may be the most scientifically advanced nation on earth, but much of its citizenry is poorly educated in this area. In the realm of genetics, we confront not simply ignorance, but obfuscation, as adherents who literally interpret the story of Genesis promulgate disinformation on the subject of human origins and deflect any rational discussion on evolution. The fervor to advance a creationist agenda in public education has spawned many lawsuits over the past century, the most recent of which was brought against a Pennsylvania school board who required students to be exposed to "intelligent design." This interview with the deciding judge in that case provides us with insight into the issues at stake.

Interviewed July 17, 2008

Published December 5, 2008

I ONCE HAD A POST-DOCTORAL FELLOW WHO, upon discovering we had grown up a quarter mile from each other in a small town in Pennsylvania, commented on our shared experience with, "Well, you know what they say about Pennsylvania? There's Philadelphia, there's Pittsburgh, and there's the state of Alabama in between." That blunt assessment (attributable to James Carville) certainly resonated when I first read about the Kitzmiller et al. v. Dover Area School District case in late 2004.

Dover, indeed, is a small town in south central Pennsylvania. At that time, the Dover school board instructed 9th grade biology teachers to read a statement that evolution is only a theory for the origin of species and to proffer an alternate explanation called "intelligent design" (ID). Tammy Kitzmiller, the mother of two students in the Dover public schools, together with a number of other plaintiffs and assisted by the American Civil Liberties Union (ACLU), sued the Dover school district for an injunction against the statement and use of materials in science class as a breach of the First Amendment of the US Constitution.

At the bench during this remarkable trial sat the Federal Judge for the Middle District of Pennsylvania, the Honorable Judge John E. Jones, III. Jones, a Republican and a Bush appointee, was assumed by many and feared by others to be inclined to rule for the defendants. However, in a stunning Memorandum Opinion (see http://www.pamd.uscourts.gov/kitzmiller/kitzmiller_342.pdf), Jones excoriated intelligent design, waxed eloquent about the meaning and practice of science, and, for the skeptics, restored faith in the fairness of the judicial system.

My call to the Judge's chambers in request for an interview was answered in vivo by his assistant, who suggested simply e-mailing the Judge directly. I did, and back came an immediate reply of "Happy to do it." On the appointed July day, in near 100-degree heat, I drove from my father's home in Pottstown along country roads through the corn-laden, cow-dotted agricultural landscape that I love. But as I got closer to my destination, the state capital of Harrisburg, billboard outcroppings disrupted the fields' quiet beauty with warnings such as, "It's your choice—heaven or hell." It appeared that I had arrived at the crux of the matter.

Gitschier: I am very excited to meet you. There are roughly three areas I want to talk to you about.

Jones: Do my best.

Gitschier: One has to do with your background—your thinking about evolution, intelligent design, creationism—going into the trial, your experience during the trial, and then afterwards—how this might have changed you.

The second is to help me through the legal stuff. I'm not a lawyer and I'm going to be writing this for an audience of geneticists.

The third is a shorter question—the ramifications of this decision on public education in the US.

So, let's first cover a little background.

Jones: I'm from Pottsville, PA, which is in the anthracite coal region of northeast central Pennsylvania. And I was raised in Orwigsburg, a little town of about 2,000 not far from there. It's in an old industrial coal county. I went to Dickinson College and Dickinson School of Law, and returned there to practice. My family roots are very deep there. It occurred to me that I'd probably be able to start a successful law practice back there and I was, happily, right about that.

Gitschier: What kind of law did you practice?

Jones: I was a general practitioner, which increasingly is a dinosaur. I used to say that I was a half-an-inch deep and a half-mile wide. I needed to know a little about a lot of different things. I was the quintessential country lawyer.

Gitschier: So—wills, small disputes?

Jones: Everything. I did a lot of litigation. I liked to go to court. I became a lawyer because of the allure of the courtroom, not necessarily to be chained to an office desk.

Gitschier: I'd like to deal with some of the legal stuff I don't understand. Kitzmiller was a suit. What does that mean? I usually think of suing for money or restitution.

Jones: That's a very good question. There is a statute, known as section 1983, in the Federal Law, and in layman's terms, it's an enabling statute and it allows you to bring suit in federal court if you believe that a constitutional right has been violated. And notably, in the context of the Dover case, it allows you to recover your fees and costs if you prevail.

When this suit was brought in December of 2004, although the statute also allows you to seek money damages, that was not the request. The request was for an injunction. An injunction is a legal ruling that stops something, typically, from happening. The plaintiffs asked for an injunction to stop the policy from being implemented in the first instance. It was to be implemented in January of 2005 after it had been enacted in 2004.

That's why it was a bench trial, and not a jury trial, to anticipate a question you may have, because when you ask for an injunction, only a judge can grant an injunction. Had they [the plaintiffs] asked for money damages, it would have been brought to a jury. They were never interested, it appeared to me, in money damages. They were interested in stopping the policy from being implemented. That was their real goal throughout the litigation.

Gitschier: You're right, I was going to ask you about why this was a bench trial.

Jones: Everybody does.

Gitschier: Because the Scopes trial [in 1925] was a jury trial.

Jones: Well, that was a criminal prosecution. John Scopes was prosecuted under a Tennessee statute, which had been little used, that prohibited the teaching of evolution.

Gitschier: Little used because nobody taught evolution in Tennessee back then! Somebody put him up to it, didn't they?

Jones: As I read it, Scopes, who was certainly pro-evolution himself, was kind of dragged into the fray and set up to teach evolution with the understanding that he would be defended. And the punishment that he was exposed to was essentially fines, so there wasn't much risk to Scopes, and of course, the benefit to Scopes was that he would be the centerpiece of this spectacular trial.

The marked difference, for historical purposes, is that Clarence Darrow, who represented Scopes, wanted to inject some scientific testimony into the trial, and the trial judge would not allow that testimony. So, it was really on the statute itself—did Scopes violate the statute itself?

Gitschier: Which he did.

Jones: He did. And the most memorable moment, as you may recall, is when Darrow called his opposing council, William Jennings Bryan, as a witness. That would never happen today. Bryan didn't have to take the stand, even then, but filled with excessive hubris, he took the stand and was eviscerated and embarrassed by Darrow. And the post-script was that Bryan died within a week of the trial.

Gitschier: I'm having trouble figuring out why we keep having this battle about fundamentalist beliefs in our public schools. I keep asking the question: Didn't we solve this problem already?

Jones: No.

Gitschier: And cutting to the chase, *have* we now solved the problem?

Jones: No.

Gitschier: OK, so let's talk through some of the background and figure out why not.

Jones: Scopes took place 80 years ago, and the matter was fairly dormant after that.

Gitschier: Why?

Jones: For decades afterwards, evolution was not substantially taught or taught at all.

Gitschier: In Tennessee or anywhere?

Jones: Anywhere. But by the '50s in the US, with Sputnik and the Cold War, there was a belief that we were falling drastically behind in science education and in other things, and you began to see a much more dedicated science component of education.

However, in certain pockets of the United States, particularly the South, there were anti-evolution statutes still on the books, and starting in the late 1960s, there was a progression of cases...

Gitschier: Starting with Epperson v. Arkansas?

Jones: Well, Susan Epperson's case. Susan was a young biology teacher who was involved in a lawsuit that had to do with a law prohibiting the teaching of evolution.

It was the same thing as Scopes, but now we're going to go after the *statute* itself. And Susan, whom I've met—a marvelous woman—was the prototypical plaintiff. She was a person of faith. She was young. She was telegenic, articulate, and she agreed to be the plaintiff in that case, which went all the way to the Supreme Court of the United States.

The result of Epperson was that a law that banned the teaching of evolution is struck down.

Gitschier: That law was in the State of Arkansas, and it was ruled that the Arkansas statute on the banning of teaching evolution was unconstitutional. Did that immediately, though, translate to other state laws?

Jones: Yes. It didn't "translate" to other state laws, but the Supreme Court, the highest court in the land, had spoken. Not per se; it ruled on the statute it had before it. But to the extent that other statutes were analogous to the Arkansas statute, the ruling meant that the wind had gone out of them. You couldn't enforce them.

Gitschier: So did it mean that evolution was now taught in Alabama or Tennessee, for example?

Jones: Not necessarily. It was still up to the school board whether they wanted to teach it or not.

But then, what states did was this: They said, "OK fine, we understand that we can't prohibit the teaching of evolution," so they developed what has been called a "balanced treatment" statute, which said that if you are going to teach evolution, then you have to teach creationism next to it.

The states said, "We must live with the Supreme Court's decision in Epperson v. Arkansas, so now we're going to try to figure a way around it and get the best deal we can. We'll hold our nose, we don't like this, but if we're going to teach evolution, we're going to teach creationism at the same time, as an alternative to evolution."

You've got a succession of cases, and I'm not trying to be encyclopedic, but again the Court said, "You're not listening. You can't teach creationism and call it what it is not."

Gitschier: These were federal court rulings?

Jones: These were all federal court rulings because they deal with the Constitution. Then following that was, "How about this, we'll have you teach 'creation science'." And after drilling into that, the Court [in Edwards v. Aguillard] said, "No, a studied examination of creation science indicates that it is nothing more than creationism labeled in a different way."

Gitschier: So, once creationism or creation science is struck down in one case, then what happens to all the other places that teach creation science?

Jones: Well, when the Supreme Court of the United States speaks, they can't do it. The bottom line is that as that line of cases concluded, you knew that you couldn't ban the teaching of evolution, you knew that you couldn't pass a "balanced treatment" statute, and you knew that you couldn't re-label creationism as creation science and have it pass constitutional muster.

Which then set the stage for intelligent design.

Gitschier: I read that you learned about this suit on the radio while driving home from work one day.

Jones: I was leaving this courthouse in Harrisburg, and I heard on the news from a local radio station that a very large lawsuit had been filed. There was a press conference at the state capitol rotunda, right across the way, by the plaintiffs' attorneys and that the suit was an establishment clause.

Gitschier: When you say "large," you don't mean financially large.

Jones: Large, meaning impactful, notable, involving a big issue. And lawyers for the plaintiffs, the ACLU and a firm from Philadelphia, Pepper Hamilton, and the plaintiffs all appeared in the capitol rotunda. And they said that the suit had been filed in the Middle District of Pennsylvania, which is my district, and I've joked since then that I had two thoughts then. One, although I consider myself reasonably well-read, I could not remember hearing about ID before, so I really didn't know what it was. And two, I wondered who would get the case. And then forgot about it until I got into my Williamsport Chambers the following morning and looked at my new cases.

Gitschier: How many Middle District judges might have seen the case?

Jones: At the time there were five of us. I got it by the luck of the draw. It rotates in a sequence. I'd like to tell you it's because I'm so good, but it was just random.

Gitschier: Tell us about your education for this case. Although you hadn't heard of ID, you likely had heard of creationism or creation science. Had this been a field that you followed at all?

Jones: No, not other than popular culture. When I went to law school in the late '70s, I followed the progression of cases that we talked about before. I understood the general theme. I'd seen *Inherit the Wind*.

Gitschier: So now it's on your docket, and you must have been curious. Did you Google intelligent design?

Jones: No. I got what I needed in the context of the case. And it was the monster on my docket.

To your question: I think laypersons apprehend that when we get a case, it's incumbent upon us to go into an intensive study mode to learn everything about it. Actually that is the wrong thing to do. The analogy is that when I have a jury trial in front of me, I always instruct jurors, particularly in this day and age when you can Google anything, not to do that. I don't want you to do any research or investigation. Everything you need to decide this case you'll get within the corners of this courtroom.

So it is with me. And I knew that by the time the case went to trial and during the trial, that I would get expert reports.

Gitschier: From whom?

Jones: Everybody. The way expert opinion works is that I get a summary of their testimony first, and that I can read in advance. So I have a flavor for it. So then the question is, why also have them testify? That is because they are subject to cross examination and everything they say may not hold up that well. And, as it turned out, some of it didn't during the trial.

In any event, I was taken to school. From the earliest point in the litigation to the time the briefs were filed, it was the equivalent of a degree in this area. Folks who disagree with my opinion will tell you I never got it right, but I'm confident that I did.

Go back to your last question. It's very critical. I have to decide cases on the facts that are before me. I can't decide a case based on my own opinion, gleaned from outside the courtroom. That's why I don't engage in my own independent investigation. If you look at other systems in other countries throughout the world, they do that. But in our system of justice in the US, we let the parties try their cases and we find the facts from what is presented to us in the courtroom. And the law, presumably we know and we apply the law. That's our job. But the facts that we apply the law to are covered at that time.

Gitschier: I don't know if you're even allowed to answer this. Before this case landed on your lap, did you have any thoughts about creationism or evolution, or the debate?

Jones: The precursor to my answer is that it doesn't matter. A judge could be an avowed creationist, but he's got to rule based on the facts and the law. In that event, he'd have to hold his nose and do his duty as a judge.

I am a person of faith. I'm certainly not an atheist or an agnostic and I see some divine force somewhere. That said, having had a pretty good education, a great liberal arts education at Dickinson College, I must say that I never had any substantial doubts about evolution generally. I had forgotten, admittedly, a lot of what I had learned about evolution back in college. Moreover, a lot had happened since the '70s, so my understanding was rudimentary. But I never had a crisis of confidence about evolution or a reason to doubt that it constituted a valid theory and good science.

Gitschier: Regarding the Memorandum Opinion itself, I found parts of it astonishing. You used words like "mendacity," "sham," "breath-taking inanity of the board's decision."

Jones: You should have been there.

Gitschier: I wish! Going into this you are impartial. What were some of the highlights? What were the transformational points in the trial that then allowed you to say, "OK, I'm going to rule this way"?

Jones: I don't think there was an epiphany. The very first witness for the plaintiffs was Ken Miller. He is very invested in this issue. He writes a textbook that is used substantially in high school biology classes throughout the country. And I think it's fair to say that the plaintiffs knew what they had in terms of their judge. They knew that I was not a scientist, but hopefully that I had a reasonably good head on my shoulders, that they were going to need to take me to school. So their first witness did just that.

I will always remember Ken Miller's testimony in the sense that he did A–Z evolution. And then got into intelligent design. And having laid the foundation with the description of evolution, got into why intelligent design doesn't work as science, to the point where it is predominantly a religious concept.

Gitschier: Is the other side objecting all the time?

Jones: They can object to a question that is truly objectionable. But there weren't a huge amount of objections. I let both parties try their case. They knew they'd have their turn.

Which gets me to the next point. Another remarkable moment on the science side was Michael Behe, who was the lead witness for the defendants, and a very amiable fellow, as was Ken Miller, but unlike Miller, in my view, Professor Behe did not distinguish himself. He did not hold up well on cross-examination.

So on the science side those were the two remarkable witnesses, although there were many expert witnesses in the field of theology, paleontology, biology, pedagogy.

Gitschier: It's almost like a command performance! There's no jury, it's not televised. All of these knowledgeable people...

Jones: Playing to an audience of one. Which was fascinating.

In the realm of the lay witnesses, if you will, some of the school board witnesses were dreadful witnesses and hence the description "breathtaking inanity" and "mendacity." In my view, they clearly lied under oath. They made a very poor account of themselves. They could not explain why they did what they did. They really didn't even know what intelligent design was. It was quite clear to me that they viewed intelligent design as a method to get creationism into the public school classroom. They were unfortunate and troublesome witnesses. Simply remarkable, in that sense.

Gitschier: Did Miller talk about molecular evolution, DNA sequences, etc.?

Jones: To the extent that he needed to.

Gitschier: Because the evidence *is* amazing.

Jones: It is stunning when you get into it. Broadly, as the trial progressed, what was remarkable to me, as you go back—you well know this in your field—people called it Darwin's theory of evolution. Here's Charles Darwin, who had not the benefit at all of genetics, and yet from my view, almost every subsequent discovery tends to bear out Darwin's theory and has only made it stronger, including the field of genetics. But Ken Miller went into the immune system, the blood clotting cascade, and the bacterial flagellum—all three are held out by intelligent design proponents as irreducibly complex, and in effect, having no precursors. He [Miller] knocked that down, I thought, quite effectively—so comprehensively and so well. By the time Miller was done testifying, over the span of a couple of days, the defendants were really already in the hole.

But I can't decide the case until I hear all the evidence, and I didn't.

Gitschier: I want to address a very specific part of your Memorandum Opinion, which is defining science. What were you trying to do here?

Jones: First of all, both sides presented ample scientific testimony, and they asked me to decide that.

Gitschier: Both parties wanted you to address the question of what is science?

Jones: Well, not what is science, but whether intelligent design is science. Why else would they have presented all those expert witnesses?

Gitschier: Do they explicitly say that?

Jones: Sure they do.

Gitschier: Is that part of the original suit?

Jones: Yes, part of the analysis—the second prong of the Lemon test and the collapsed endorsement test [see Sidebar]—is the effect on the intended recipients. My view, and I'll always believe that I was right about this until somebody convinces me otherwise, is that if you're going to measure the effect of a particular policy, in this case juxtaposing intelligent design with evolution, on the intended recipients, you have to delve into what the policy is about. What was it about? It was about intelligent design. And to try to determine the effect on the recipients you have to determine what does that concept or phrase stand for? Hence, we got into a search and examination of what exactly does ID say, what is its basis, what are its scientific bona fides or lack thereof. That opens the door for a determination of whether ID is in fact science. And that is what that part of the opinion was.

THE JUDGE PROVIDES A PRIMER

Gitschier: There are a number of things in your Memorandum Opinion that I want you to flesh out for our readers. One is the Establishment Clause of the First Amendment. Second is the Lemon test and the prongs of the Lemon test. And the third thing that I really wasn't clear on was the endorsement test.

Jones: Lot of lawyers aren't clear on that either; it's very complex.

The Establishment Clause as contained in the First Amendment, simply stated, says that Congress shall pass no law that, in effect, favors an established religion. It's been the subject of a great deal of debate. Initially, in its inception, it was applicable only to Federal government, but with the Fourteenth Amendment, it was made applicable to the states, and hence, applicable to any governmental or quasi-governmental body including a school board. So there is no debate that the school board was subject to strictures of the Establishment Clause of the First Amendment.

There is a vigorous debate that takes place, to this day, as to whether there is a wall of separation between church and state, as Thomas Jefferson opined. That phrase doesn't appear anywhere in the Constitution. However, the Supreme Court of the United States has clearly set out, in its decisions over the past 60 years, that there is a wall, porous at times, but a wall nonetheless. So the common theme of their decisions is that they are going to look with a high degree of scrutiny on government activity that seems to favor a particular religious concept. Hence the line of cases we talked about before.

Now the devil is in the details, and so it then fell to the Justices to develop, as they typically do in cases like this, tests—overlays, if you will—that they put against the facts that are found by judges, so that the judges can decide whether a violation has occurred. As you might expect, because every case is so intensively fact-specific, sometimes these tests are really hard to apply.

So the first test that the Court came up with is the Lemon test, Lemon v. Kurtzman [another Pennsylvania case regarding the reimbursement of Catholic schools by the state superintendent of schools].

What came out of Lemon were three prongs that judges have to look at. The first is: what is the purpose of the enactment? The second is: what is the effect of the enactment? And the third is: is there an excessive entanglement between religion and government?

I'll come back and be specific to my case [in a minute]. As time went by, it was apparent that the Lemon test was somewhat difficult to apply in certain factual situations. In particular it was found to be difficult to apply in cases where, for example, the Ten Commandments were bolted onto the side of a courthouse or government building. So Former Justice Sandra Day O'Connor then penned the "endorsement test." The endorsement test, boiled down to its essence, takes the first two prongs—the purpose and the effect prongs—and collapses them together, and just makes it easier to apply, although it is always hard to judge these cases.

To go back to the Lemon test. If the judge finds that the purpose is predominantly religious, you can stop; you don't have to go to the other prongs. But if you find it's OK, you can go to the effect prong—what is the effect on the intended recipients of the policy? How do they view it? If you find a violation there, you needn't go to the excessive entanglement prong.

In my case [Kitzmiller], it failed the purpose prong, and the excessive entanglement prong was never at issue, by agreement of counsel on both sides. But for the sake of completeness, because I had to believe that my decision would be appealed, I did the effect prong as well. And I also did the endorsement test. But the endorsement test is just a variation on the Lemon test, and is in some ways a duplication of the Lemon test, with a twist.

People shouldn't mischaracterize it and say that I am the arbiter of what science is broadly. It's not what I wrote about in the opinion. I wrote about whether ID, as presented to me, in that courtroom from September to November of 2005, was science, and I said it was not. That it was the progeny, the successor to creationism and creation science. That it was dressed-up creationism.

Gitschier: Nonetheless, you have captured the essence of science in your opinion.

Jones: Well, you could read it that way if you chose to. What it does contain is something that you could utilize as a portable mechanism to look at other concepts and decide whether they were science. But the question I decided was whether ID was science. And you use tools like—is it testable? Is it peer reviewed? Is it generally accepted in the scientific community? And the answer to all three of those things is "No."

Gitschier: Let's talk about what happened downstream of this decision. How will this change affect the landscape of education in the US?

Jones: The short answer is that I don't know. In the two and a half years since the opinion was released, no one has tried to teach ID in the US. Remember, the opinion doesn't have precedential effect outside of Pennsylvania. In other words, I am a Federal District court with jurisdiction over this big middle of Pennsylvania, but I'm not the Supreme Court of the United States. So, it's unlike the mandates from the Supreme Court that we were discussing earlier such as Epperson and Edwards. Those are the final words for now, and everyone must adhere to them. I suppose a school board in another state could still pass a law mandating the teaching of ID, and in fact some were considering doing so at the time of this trial, but later pulled them down. But I do think that many consider my opinion persuasive, if not binding, and that's why you have not seen these policies enacted.

Gitschier: Such as in Kansas?

Jones: Such as in Kansas. Kansas at that time was having [state-wide] school board elections. And this became an issue in Kansas, and Kansans did not elect proponents of ID, utilizing my decision I think, saying that it was improvident to do this. In Ohio, they had begun steps that would have allowed the teaching of ID, and the school board ruled the policy back because of my decision, not because they had to, but they thought it was persuasive. Florida had a debate last year, into this year about changing some of their standards or adopting new standards of science, again citing my decision.

The hotbeds today—and this is re-emerging—Texas has a very strong desire to get into something like teaching intelligent design. Louisiana just passed a stature that seems like it could be used as a vehicle for teaching ID. This is speculation on my part—I don't think that the concept of ID itself has a lot of vitality going forward. The Dover trial discredited that thing that is ID. To the extent that I follow it—I'm curious about it, but it doesn't go any further than that—the likely tack going forward is something like teach the *controversy*, talk about the alleged flaws and gaps in the theory of evolution and go to that place first.

They gave me the last word in "Judgment Day" [a NOVA program on the trial] and I said this is not something that will be settled in my time or even in my grandchildren's lifetimes. It's an enduring, quintessentially American, dispute. If you poll in the US today, you'll find that approximately half of our fellow citizens believe in creationism and think that creationism ought to be taught.

Gitschier: I had no idea!

Jones: Believe me. Remember, the Dover School Board was comprised of young-earth creationists. They believe that the Bible is the Word. They either can't explain or like not to explain the evidence to the contrary. Then there are the mixed-bag creationists—creationists who accept that the world is as old as it is but don't accept evolutionary mechanism.

Gitschier: How has this trial changed your life? Both externally and in the way you think about the world.

Jones: It's changed my life forever. You can't go through something like this that has such notoriety without being changed. Federal Judges at any level lead quite cloistered existences, and I was thrust onto the stage in a way that I would never have thought possible. And I have been speaking all around the US, but I don't go and try to say what I did in the opinion.

What I developed was a passion for the concept known as "judicial independence," meaning that concomitantly with the science education issue that I just raised, I don't think Americans understand how judges operate.

I had a lot of criticism after this decision; a lot, I think, was born out of ignorance about how we do things. People didn't understand there was a Lemon test or an Endorsement test. People thought I made this up as I went along. They think judges rule according to their own philosophical biases or predilections. I thought it was incumbent upon me to get out and talk about that and say, "Well, you don't quite have this right," and I've been very well received across the country.

But from the NOVA show to the now four books that have been written about the case, to being on the cover of Time magazine, for someone born and raised in a town of 2,000 in upstate PA—all this is fairly miraculous stuff that I never thought I would do. So, it certainly has changed the fabric of my life, that I have had this interval. It will die down, I know.

When I take my last breath and they publish my obituary, the first line will say that I presided over the intelligent design trial. I can't *top* this, I don't think, and I'm fine with that, if this is what I'm remembered for. I'm proud of what I did. I thought I discharged my obligations and my duties well.

Going forward, has it made me more curious about the issue? Yes, and I think I'll always have that enduring curiosity.

Recommended reading from the judge:

Summer for the Gods by Edward J. Larson
The Devil in Dover by Lauri Lebo
40 Days and 40 Nights by Matthew Chapman

The Gift of Observation

An Interview with Mary Lyon

In humans and other mammals, sex is determined by the presence of the Y sex chromosome, which contains a small number of genes important for male sexual development. In contrast, the X chromosome is present in two copies in females and one copy in males. Because the X chromosome contains hundreds of genes that are important for human development and health, parity must be achieved between males and females so that equal levels of gene products derived from the X chromosome are amassed in the two sexes. Fifty years ago, Mary Lyon generated an idea for how this balance is effected in mammals, that during female embryogenesis, one of the two X-chromosomes is randomly "inactivated" in each cell, i.e., that its genes are irreversibly turned off. In this interview, Lyon reflects on her long career and this insight in particular.

Interviewed June 23, 2009

Published January 22, 2010

FOR MORE THAN 60 YEARS, Mary Lyon has had an intimate relationship with the house mouse. She has devoted herself to the discovery and description of a wide variety of mutants, arguably as prolific as anyone in the field. She co-edited the mouse bible "Genetic Variants and Strains of the Laboratory Mouse" and untangled the knots in the t-complex. And, in a link with posterity, her last name now forms the basis for a word—"lyonization"—synonymous with the mammalian random X-inactivation process that she first hypothesized a half-century ago.

I was keen to interview Mary but hesitant, as I knew she had retired. Thanks go to my fellow *PLoS Genetics* editor, Elizabeth [Lizzy] Fisher, who had done graduate work with Lyon and encouraged me to e-mail her. While no longer running a lab, Lyon still comes to work a few days a week at the Medical Research Council (MRC) Unit at Harwell in the United Kingdom, and she agreed to meet with me there.

En route to see Mary, as she seems to be universally and reverentially referenced, I found Harwell itself a study in contrasts. The facility surprised me in its starkness, its aging buildings and dandelion-bespeckled grass surrounded by a chain-link fence, apparently in response to the potential threat by animal rights activists. I struck up a conversation with the guard, who opined that Mary has been unfairly denied a knighthood, not only because she is a woman, but also, perhaps, because work involving animals is politically charged. I then wended my way toward the meeting room and was ushered in. The door opened to the warmth of Mary standing there, wooden cane in hand, radiating a smile, and quietly waiting to offer me a beverage. Despite her soft voice, I knew I was in the presence of a giant.

***

Lyon: Would you like a cup of tea or coffee?

Gitschier: Yes. Would you? Tea?

I want to thank you for agreeing to be interviewed. I'm interested first in your upbringing and what got you interested in science.

Lyon: I was the eldest of three children. My father was a civil servant; my mother was a school teacher when she was young.

My family lived in several places. I was born in Norwich, and then my parents moved to Yorkshire when I was four to six, then to Birmingham when I was 10 and then to Woking in Surrey when I was 14.

The grammar school I attended in Birmingham was a very good school. I got interested in science there. At first I was interested in physics and chemistry but then I quickly changed to biology. I won a prize in an essay competition when I was about age ten, and the prizes were four books on nature study. And that got me interested in biology.

Gitschier: What about your brother and sister? Were they also interested in science?

Lyon: No, they weren't. My brother became an accountant. My sister first worked as a school teacher and then as a social worker.

But the person who was interested in science is my father's sister's son, Kenneth Blaxter. He was an expert on farm animal nutrition. He was the director of the Rowett Nutrition Research Laboratory in Scotland. He won prizes and he was knighted and so on.

Gitschier: For university, you chose to attend Cambridge. Was it very common for women to be in Cambridge at the time?

Lyon: No. At that time women were not members of the University. There were two colleges for women. I was in Girton and the other was Newnham, but the women were in the minority because they were restricted to these two colleges. The men restricted us to 500 women, and there were more than 5,000 men. We used to go to the lectures with men, took the same practical courses as the men, and took the same exams as the men, but, officially, we didn't get a degree. We got a "titular" degree.

Gitschier: Really? When did you graduate?

Lyon: I graduated in 1946. And of course, the Second World War changed the position of women in the world. And in 1948, Cambridge admitted women [officially] to the University.

Gitschier: It must have been very unusual for women to go on to do a PhD.

Lyon: Yes, it was. I was in a women's college, of course, and several women went on to a PhD.

Gitschier: What did your parents think about your choice to continue with a PhD? Were they supportive?

Lyon: Yes, I think so. They wanted me to get married at one point.

Gitschier: What did you think about that idea?

Lyon: I didn't like it.

Gitschier: Was there someone in particular they wanted you to marry?

Lyon: No.

Gitschier: Just in principle, then! So, because of the war, there seemed to be more educational opportunities for you.

Lyon: Yes, there were. I didn't really realize how much more opportunity there was, but there certainly was at the time. This was because during the war, the government restricted very much the men who could go to university.

Medical students could go to university and men doing physics and chemistry, because physics and chemistry were needed in the war effort. But in zoology there were very few men. The government did allow a few men to do things like zoology and botany if they were really good, believing that if they got high marks on their exams they would do some research that would help the war effort.

But there were fewer men in relation to the number of women at that time because they were called up to the military.

Gitschier: When you graduated in 1946, [C.H.] Waddington wasn't there, but you had wanted to work with him. You ended up with [R.A.] Fisher instead. What was he like?

Lyon: Fisher was a very brilliant man but a very eccentric man. He was difficult to work with. He was brilliant in a logical mathematical sense. So we learned about ratios of normal and affected offspring, that sort of thing.

Gitschier: How old was Fisher when you worked with him?

Lyon: In his 50s.

Gitschier: Had he worked with mice for a very long time?

Lyon: Yes, I think so. He was appointed Professor of Genetics in Cambridge round about the time I came to Cambridge in 1943. Before that he was in the Rothamsted plant research station in the outskirts of London. In Cambridge he worked on both plants and mice.

Gitschier: And did he have other graduate students besides you?

Lyon: Sir Walter Bodmer was a student of his [post-Lyon]. Anthony Edwards was also a student, a contemporary of Walter Bodmer, I think. Various people went to work at the Cambridge lab: Douglas Falconer, Toby Carter. But Fisher didn't get along very easily with people, and he threw out most of them.

Gitschier: But he didn't throw you out!

Lyon: Well, I felt that I didn't have enough facilities to do my PhD there. I was trying to do dissection of mice and to breed mice and needed facilities for histology. So I moved to Edinburgh, which is where Waddington was, in Genetics.

And there were facilities for doing mouse genetics. Douglas Falconer was my supervisor there in Edinburgh.

Gitschier: I see. So, you all jumped ship! What became of poor Fisher?

Lyon: He did have people who worked with him for short times and there were one or two people who did get on with him and stayed there. These included Margaret Wallace and George Owen. But Fisher stayed there until he was retiring age.

Gitschier: So you started to work on the *pallid* mutant when you were still with Fisher and continued on with that project for your thesis. When did you finish your PhD?

Lyon: 1950.

Gitschier: At that point what was happening?

Lyon: Waddington was very good about getting money for young people to stay on at Edinburgh. He sent in an application to the MRC for a project for me to do. He didn't think of me working for the ARC [Agricultural Research Council] because they weren't giving equal pay for women at the time. He got the MRC money, and that's how I started my post-doc in Edinburgh.

Gitschier: And what did you work on?

Lyon: I continued to work on *pallid*, but I also worked on the mutagenic effects of radiation, part of a bigger project that Waddington got the money to work on, namely, the mutagenic effect. At that time, after the Second World War and the atomic bombs in Japan, there was a lot of concern about the harmful effects of fallout in the atmosphere. So I was part of this project, which also included studying the actual mutants that we had obtained.

Gitschier: What kind of mutants?

Lyon: I will just mention a few examples. One was called *ataxia*, a mutant of the nervous system that caused the mouse to have problems walking. There was also *twirler*, a mutant that affected the inner ear of mice—they ran around in circles and had no sense of balance and shook their heads. There were *short-eared* mutants and a type of vitamin-D resistant rickets.

Gitschier: A wide spectrum of things!

Lyon: The [mutagenesis] project there got us all scared. The head person responsible for the experiment, Toby Carter, said that he couldn't do the mutagenesis experiment without a lot more breeding cages for the mice. And there was no possibility of getting these extra facilities in Edinburgh. Toby was in contact with John Loutit, who was the director of this unit here [in Harwell].

Gitschier: So Harwell already existed.

Lyon: The MRC had this project at Harwell to study the harmful effects of radiation.

Gitschier: So, similar to the Edinburgh project.

Lyon: Yes, but they were not doing genetics, they were studying cancer [in mice].

Gitschier: There is another woman whose name is on a lot of the papers with you at this time—Rita Phillips. Who is she?

Lyon: She was employed as a research assistant in Edinburgh before I was there. And she came to Harwell, too. We moved in 1955.

Gitschier: So did you continue to be a post-doc upon your move to Harwell?

Lyon: No. People didn't talk about post-docs in those days. People were scientists. You could have a short contract or you could have tenure.

Gitschier: So you were a scientist, presumably with a short contract. Renewable?

Lyon: Yes. First I had a 3-year contract, then a 5-year contract, then tenure.

Gitschier: I want to talk to you about the new X-linked mutants, such as *Tabby* and *mottled*, that were starting to be identified. Where were they discovered?

Lyon: They were discovered in Edinburgh by people working with Douglas Falconer.

Gitschier: So, even before you moved here, you knew about these mutants.

Lyon: Yes, it was a very exciting thing to talk about in those days. No one had found a sex-linked mutant in mouse until then. But we didn't pursue it initially.

Gitschier: Did the mutants move to Harwell too?

Lyon: Yes, there were quite a lot of different *mottled* mutants early on, and we didn't have all of them. There aren't so many *tabby* mutants, but we did have *Tabby*.

Gitschier: When did you first start having this idea about the X chromosome inactivating?

Lyon: I was still studying the mutants that we had found in mutagenesis experiments. We found quite a number of *mottled* [mutants], and they weren't all the same. In some the affected males die as embryos; in others they are born and have white coats. The females were variegated. And I found one in which the original animal of this particular mutant was a mottled male, which was odd because males have got only one X chromosome. So why was he mottled?

So we bred from it to find whether the mottled pattern was inherited. This mouse had some daughters who looked like himself and he also had normal daughters and normal sons. So his mottled appearance was inherited. When we bred from his affected daughters, they bred as the previous *mottled* mutants that had been found. That is they had mottled daughters, like themselves, and also affected males, which died. So the females were behaving like ordinary *mottled* mice with a mutant gene on their X chromosome.

But we still had the question of the original mottled male mouse. How did he get to be mottled? Then it occurred to me that he had a mutation that had occurred in him, when he was just an embryo, when he was just a few cells, and that gave rise to one progeny group of cells with a mutant X chromosome and another group of cells with the unmutated, normal X chromosome. So this original mutant male was a mosaic of two types of cells, some with the mutated X chromosome and others with the normal X chromosome.

So then, it occurred to me that if that explanation of him having two types of cells applied to his pattern, could it not also apply to the pattern of his daughters? His daughters could have two types of cells, one with the mutant gene active and one with the normal gene active.

And that involved me in finding out about recent work on the mammalian X chromosome. One important point was that XO mice are normal fertile females, and thus a female mouse needs only one X chromosome for normal development. Furthermore, female mammals have the sex chromatin in their nuclei, and, just recently before that time, Ohno had found that the sex chromatin consisted of one highly condensed X chromosome.

So the female mouse only needs one X chromosome, and in female mice the X chromosome behaves strangely. So I put all those things together and came up with the idea of X-chromosome inactivation.

Gitschier: Before you read the literature and pieced all this together, did you already have the idea that in females only one X was active?

Lyon: Yes.

Gitschier: These mice that you are referring to: were they also the product of radiation?

Lyon: No, the original male was a spontaneous mutant.

Gitschier: Do you remember the year that original male appeared?

Lyon: 1959 or 1960.

Gitschier: You published your paper in 1961, so the pieces of the puzzle must have very quickly fallen into place. And do I take it that X-inactivation is also playing a role in the *Tabby* mutant?

Lyon: The striped pattern in *Tabby* females is indeed due to X-inactivation. It is not due to differences in pigmentation of the coat but to differences in hair texture. *Tabby* males have an obviously abnormal coat, which looks too sleek. Females have patches of this abnormal hair and where the patches of mutant and normal hair meet, one sees a stripe. The sizes and shapes of patches and stripes in heterozygotes for different X-linked genes depend on the way that the cells underlying the patches migrate and mingle during development.

An interesting example concerns the tortoiseshell cat. The pattern is produced by cells giving black or yellow pigment. If the cat has an autosomal gene for white spotting, patches of black and yellow are larger. This is because the spotting gene reduces the number of pigment cells and hence each precursor cell must cover a wider area and hence produces a larger patch.

Gitschier: Lizzy Fisher mentioned to me that one person in particular, Hans Grüneberg, gave you a lot of grief about your hypothesis. Would you like to comment on that and whether that was difficult for you?

Lyon: Grüneberg did indeed make things difficult in the early days of X-inactivation. He seemed to have two main objections. Firstly he seemed to think that I was not sufficiently established or eminent enough to put forward such a major idea. Secondly he seemed to have problems with the points mentioned above on sizes and shapes of stripes and patches. The theory does not require that each stripe or patch be derived from a single precursor cell. The *tabby* gene in the mouse provides an example. The gene affects the development of the teeth. If each tooth were derived from a single precursor cell, then each one would be either fully mutant in phenotype or fully normal. In fact, each tooth is intermediate in appearance. This is consistent with the origin of each tooth formed by a small pool of precursors in which some cells have the mutant gene and others the normal gene active. Individual teeth will vary in the proportion of precursor cells with the mutant gene active. Grüneberg seemed to find this difficult. His objections made it difficult to study the stripes and patches of heterozygotes, which were an important source of information in the early days before molecular methods were available.

Gitschier: You have now become interested in a new hypothesis, that LINE elements on the X chromosome can serve as a means of transmitting X-inactivation in *cis*. That they are somehow boosters. How did you come up with this hypothesis, and are you alone in this theory?

Lyon: I thought of this a long time ago because some of the early work on the mouse X chromosome involved X-autosome translocations. And the autosomal part of the translocation does not get inactivated as efficiently as the X chromosome does. Similar evidence from other translocations suggested to me that X-inactivation travels less well in autosomes than it does in X-chromosome material.

So how could that happen? What could there be in X chromosomes that facilitates the spread? I thought it would be something promoting the spread in X chromosomes, rather than inhibiting the spread in autosomes. What could it be? There is a limit to what it could be.

Gitschier: What were some of the things you ruled out?

Lyon: *Drosophila* have the *roX* genes that work in dosage compensation. But in mammals no one had ever found anything like that. Mammals had not evolved that kind of gene. So what could it be that served as a boosting agent?

And I thought of repetitive elements as booster elements several years before the LINE hypothesis came out. People have found that the X chromosome of the human, and I think also the mouse, is particularly rich in LINE elements, compared to the autosomes. So I thought it could be repetitive elements, particularly the LINE elements.

Since then, there are even more data in the literature to support this, data that come from the human genome sequencing project. The human X chromosome is very rich in LINE elements, particularly in regions where most genes are inactivated, whereas the regions of the X where inactivation does not occur very efficiently are not rich in LINE elements.

But there are other bits of evidence that have not supported the LINE idea terribly well.

Gitschier: What kind of evidence?

Lyon: There are some odd animals, odd species, that have different types of X-inactivation and weird types of DNA, in which LINE elements are not terribly active—not alive—not transcribed. There are some species of vertebrates that have no active LINE elements, but that have X-inactivation.

Gitschier: What kind of species?

Lyon: Particular species of wild mice and rats.

Gitschier: Well, is the fact that they be actively transcribed a necessary part of your hypothesis?

Lyon: No, I don't think it is.

Gitschier: Well then, mechanistically, what could it be about the LINE elements that could make them boosters?

Lyon: That is still to be found out.

Gitschier: It will be interesting to watch this story evolve.

You wrote a personal history of a half century of mouse work in which you comment that this is just the "hors d'oeuvres and the feast is yet to come." So I'm wondering if you were to be able to start all over today, is there a project you would choose to work on? I assume you would still choose to be a mouse geneticist!

Lyon: I think so, yes. It would be nice to work on the genetics of behavior. This is an area that will be interesting to work out.

Gitschier: Did you feel a life in research was a good fit for you?

Lyon: I think so, yes. Teasing out problems and applying the scientific method to problems. The thing I didn't like about it when I got to retirement age was how much admin there is: staff appraisals, annual reports, project costings. And there is a lot of admin to do with animal experiments in this country.

Gitschier: When did you retire?

Lyon: 1990.

Gitschier: Were you required to retire [because of age]?

Lyon: Yes.

Gitschier: Do you have a cat?

Lyon: Yes!

Gitschier: Me too! What's your cat's name?

Lyon: Cindy.

Gitschier: And you have a building named after you now. Was that a surprise?

Lyon: Yes it was!

Gitschier: When you look back on your scientific career, what did you enjoy the most?

Lyon: The time I spent in Edinburgh, I would say. It was a very lively academic atmosphere. Leaving and coming here wasn't very good, because we left a big genetics lab and a lot of able and enthusiastic geneticists. Here, there were hardly any other geneticists, and the people weren't as enthusiastic. [But] things did improve here.

Gitschier: Are there any other topics you would like to talk about? As long as it's not the *t*-complex, I'm OK.

Lyon: People always say about the *t*-complex that they can't understand it. But it seems very sensible to me.

Gitschier: I think I'd have to warm up to it. Perhaps some more tea?

Imagine

An Interview with Svante Pääbo

Within the past 40,000 years, two human species co-inhabited Europe: ours and the Neanderthal, Homo neanderthalensis. *The Neanderthal, like other archaic human species such as* Homo erectus, *died out for unknown reasons and was replaced by modern humans throughout the world. This interview conveys one person's quest for understanding the genetic relationship between* Homo sapiens *and* Homo neanderthalensis *and the rigors of working with ancient DNA specimens.*

Interviewed October 26, 2007

Published March 28, 2008

154 S. Pääbo

SVANTE PÄÄBO WORKS ON THE EDGE OF WHAT'S POSSIBLE. He ignites our imagination, unlocking tightly held secrets in ancient remains. By patiently and meticulously working out techniques to extract genetic information from skin, teeth, bones, and excrement, Pääbo has become the leader of the ancient DNA pack. Sloths, cave bears, moas, wooly mammoths, extinct bees, and Neanderthals—all have succumbed to his scrutiny.

Pääbo broke ground in 1985, working surreptitiously at night in the lab where he conducted his unrelated PhD research, to extract, clone, and sequence DNA from an Egyptian mummy. From there, he joined the late Allan Wilson as a post-doctoral fellow in Berkeley, where together they rejuvenated sequences from extinct species. Returning to Europe, he landed a full professor position in Munich. He is now Director of Evolutionary Genetics at the Max Planck Institute for Evolutionary Anthropology in Leipzig.

A sterile hotel lobby wasn't the venue I had hoped for in interviewing Pääbo. I would have preferred a natural history museum, or, even better, an archaeological dig to stimulate the interview juices. But, when I realized he was attending the American Society of Human Genetics meeting in San Diego last fall, I grabbed the opportunity. Though jet-lagged, he gamely agreed to a 10 p.m. interview, following the Presidential symposium and only seven hours prior to his planned surfing excursion in La Jolla.

Gitschier: What happened in your youth to make you so interested in Egypt?

Pääbo: Sometime in my late boyhood, I got very interested in archeology. I went around after big storms in Sweden to spots in which trees had fallen over. You can look at the roots for things—stone age pottery and things like that. Even in the suburbs of Stockholm, where I grew up, there was still a forest around. And you could run around and have fun. It certainly was common for kids to play "stone age" behind the school in the forest.

Gitschier: Was there something that triggered your particular interest in archeology?

Pääbo: Not really, but I think it was the realization that you *could* actually go out yourself and find these things!

Gitschier: And did you find stuff?

Pääbo: Yes, they are still at my mother's place, in a glass cabinet—thousands of pot shards that I collected. You can sometimes passel them together and can get part of a pot that was used 3,000 years ago. Quite fascinating.

Also, my mother had taken me to Egypt because I was interested in Egyptology. I think I was 14. That made me fascinated, as so many young kids are, with Egypt and mummies and pyramids. It was mainly the trips I took to Egypt—three times with my mom.

Gitschier: Wow, was your mother into Egypt, too?

Pääbo: It was partially through my fascination, but I think she still goes to lectures on Egyptology in Stockholm.

Gitschier: Were your parents scientists?

Pääbo: Yes. I grew up with my mother. My mom and dad were not married. My mom was a chemist and worked in industry. My dad had another family, but he was a biochemist and studied prostaglandins.

Gitschier: And then you worked in biochemistry?

Pääbo: I first started studying Egyptology and things like that at the University [Uppsala] and got somehow disappointed. It was not as romantic as I thought it would be. And after a year and a half or so, I didn't know what to do, because this wasn't really "it". So I started studying medicine because I figured I would get a profession. And it was also a way into basic research.

Gitschier: I read your paper from 1985 about sequencing the mummy remains. What was the genesis of that?

Pääbo: I knew there were hundreds and thousands of mummies around in museums and that they found hundreds of new ones every year, and molecular cloning in bacteria was a rather new thing at the time, so I found in the literature that no one had tried to extract DNA from Egyptian mummies, or any old remains actually. So I started to do that as a hobby in late evenings and weekends, secretly from my thesis advisor.

Gitschier: As a lowly graduate student, where do you find a piece of mummy to start this investigation?

Pääbo: I had studied Egyptology, so the professor of Egyptology knew me quite well. He helped me to sample a mummy in the museum in Uppsala. He also had very good connections with a very large museum in Berlin, which was East Berlin at the time. Germany has a long, long tradition in Egyptology, going back to the 19th century. After the British Museum and the Museum in Paris, the Berlin Museum has the biggest collection outside Egypt.

Gitschier: So you went with your professor to the museum in East Berlin. . .

Pääbo: He had convinced them of our idea in advance. We sampled, I think, 36 different mummies. Small samples, of course.

Gitschier: Had people ever looked at mummy tissue before, at things like proteins?

Pääbo: There had been some work on histology of mummies, and there had been some work on trying immunoreactivity of proteins extracted from it, with very mixed results. I don't think there were any convincing results from Egyptian mummies.

Gitschier: In what kind of state are the mummies? Are you wearing gloves or masks? What are you doing?

Pääbo: We only worked with mummies that were already unwrapped and with things that were broken, so we were not destroying anything to get to the tissues. With a scalpel we removed a little piece. It was the first time this was done, so we had no big qualms about contamination. I had no idea this could be such a big issue.

Gitschier: What did you do with these 36 scalpeled samples?

Pääbo: We screened them with histology. We looked at both with traditional stainings—hematoxylin-and-eosin staining and staining with ethidium bromide—and under UV light to see if one could see any fluorescence from DNA. In the skin of a particular mummy, you could see that the cell nuclei lit up. So, there was DNA there and at the place you would expect it to be.

Gitschier: Was your interest in this simply the challenge of getting DNA sequence out of it or was there a bigger idea?

Pääbo: It was clearly the idea that if you could study the DNA of ancient Egyptians, you could elucidate aspects of Egyptian history that you couldn't by traditional sources of archeology and the written records.

Gitschier: Do you mean the relationships between people?

Pääbo: Population history. Say, when Alexander the Great conquered Egypt, did that mean there were lots of people from Greece who actually came there and settled there? When the Assyrians came there, did that have an influence? Or was the population continuous? Political things that influenced the population.

Since then it has become clear that it is almost impossible to work with human remains because of contamination. It is very hard to exclude that the DNA you look at is not contaminated with modern humans.

Gitschier: Then, how do we know that this sequence in the 1985 paper is in fact the sequence of a real Egyptian?

Pääbo: In hindsight, we don't know that. In 1985, I had no idea how hard this is [to retrieve uncontaminated ancient DNA sequences] and thus did not do the controls we now know are necessary. We've even published at a later point on this.

Gitschier: But there have been no data to refute the sequence of this mummy.

Pääbo: But nothing to prove it either! It could well have been contamination, and if that was the last that had ever been written on ancient DNA, that would have been a sad state of affairs and the end of the field.

Gitschier: Have people gone on to look at more mummy DNA since then?

Pääbo: Egyptian mummies are actually quite badly preserved; also animal mummies. This probably has to do with climate. It seems the cooler it is, the better preserved things are. We looked at a few Neanderthal remains from Israel and Palestine and they have so far not yielded any DNA.

Gitschier: What is it like to travel all over the world to try to get specimens?

Pääbo: To sample these things takes building confidence—in museum curators and archeologists and paleontologists—that we *can* actually get information from them. And, of course, it is a balance for a curator between a destructive sampling for scientific progress against responsibility for future generations to preserve these things. With justification, you can sometimes say that if you can just wait 30 years, methods will be so much better.

What you actually do is a several stage process, where you first take very small samples, of say 10 mg, and just see if there are amino acids preserved—the amino acid profile of collagen. If there is no collagen preservation, it turns out there is hardly ever DNA. We can already exclude a lot of remains that way.

And then we take samples of 100–200 mg, extract DNA, and see if we can find Neanderthal DNA. And then for the genome project where we need larger samples, we use bones that have very little morphological information. So in the Museum in Zagreb, which houses the Vindija remains, we screen bones of which it cannot be said from the morphology if they are human or animal. By doing extraction from 100 mg, you can determine the species from the mitochondrial DNA. So the paleontologists gain something—they learn what species the different bones come from, and so it is easy to justify taking half a gram from them if they turn out to be Neanderthal bones.

Gitschier: So now, back to the mummies. That was not your thesis.

Pääbo: No, but then I had to tell my thesis advisor that I had done this! He was happy that it had been successful. I don't think he would have been so happy if I had presented it before it happened.

Gitschier: Then you went to Allan Wilson's lab. What did you work on there?

Pääbo: Really developing the technology for ancient DNA. PCR had just come around, and I had tried to do PCR back in Europe with water baths. It was really when Taq polymerase came and the thermocycler, and Allan's lab was, I think, the first academic lab to have one.

Gitschier: And what organism were you working on?

Pääbo: We started again on Egyptian mummies but rather soon switched to animals of different sorts. We worked on moas from New Zealand, which are extinct flightless birds, ground sloths, and the marsupial wolf from Australia.

Gitschier: Why did you switch? Better preserved or no contamination issues?

Pääbo: Contamination with human DNA became apparent rather soon when I started using PCR, when you could repeat experiments and do lots of negative controls.

Gitschier: Was your passion, though, in these extinct animals? Or was your intent to go back to human lineages?

Pääbo: It took on its own life. It became very fascinating to develop the technology and overcome the problems with contamination, problems with errors in the sequences, things like that.

Gitschier: How did your move to Germany come about?

Pääbo: Pretty much by chance. I had a girlfriend—I've had both boyfriends and girlfriends in my life—but at some point I had a girlfriend who was from Munich. The professor of genetics there asked me to give a seminar at some point, and then he said they had this professorship coming up in a year and I ought to apply, which I did, and by the time it had all worked out ... I had no girlfriend there anymore!

But, it was clearly a very, very good offer. The biggest break I got in my life. I became a full professor there after being a post-doc, directly, without being an assistant professor. The opposite to your prejudice about how European science works, in that case. There was a constellation of people there who were not risk-averse.

Gitschier: How did the Institute in Leipzig come to be?

Pääbo: After German reunification, there was the political will and the money to start institutes in East Germany at the same density, according to population, as there were in West Germany. This was the chance to start a number of new

Max Planck Institutes. And there was a very conscious idea to ask—in what areas of science is Germany particularly weak? And of course, anthropology is such.

Gitschier: They were weak in anthropology?

Pääbo: Absolutely. Due to what happened during Nazi times. There had been an Institute of Anthropology where Mengele was an assistant. And so no one had really wanted to touch anthropology since that time.

So there was a lot discussion if one would dare to do it or if it was too politically sensitive. And then, once the decision was made to actually do something in the direction of human evolution, it was in fact a big advantage that there were no big traditions. Because you could say—how would we *now* start an institute in evolutionary anthropology not burdened by any traditions? And the idea sort of grew among people who discussed this. If we were to do this, we would ask the question of what makes humans unique in a comparative way across different disciplines—humanities or sciences—but it should all be empirical, not just a question of philosophy.

It ended up being an institute with five departments: Paleontology; Primatology, with research sites in Africa, studying chimps and gorillas in their natural habitats and their range of behaviors; Comparative Psychology, which has a primate facility in Leipzig, the only research facility in the world with all the great apes, and it's part of the zoo. Visitors can actually observe the experiments. They do experiments in cognitive development in human children and ape children for the first 12 months of life—the very same experiments. When do you see the things that set humans apart and what are these things? And there is Comparative Linguistics—what is common to all human languages? And then Genetics.

Gitschier: What have you got your eye on, other than Neanderthals?

Pääbo: We are very interested in comparative genomics of the apes in general. We are sequencing the bonobo—the last ape that has not been sequenced—with the 454 technology. The amazing thing is with these high-throughput technologies, a lab can now take on projects that a genome center did just a few years ago.

For the Neanderthal, we have to do so much sequencing that we do it with the 454 company in a collaboration. We test the libraries we make in the clean room, and when we have a good library, it goes to Bradford [Connecticut] for the production sequencing.

Gitschier: Do people come to you with crazy ideas that intrigue you?

Pääbo: Yes. Our lab pretty much functions on the ideas that are born in the group, and our lab is a little unusual in that we spend a lot of time discussing every project every week, in a group setting. All the groups that have to do with ancient DNA—all the people sit together once a week and discuss their work—particularly things that don't work. Gene expression is another day, or genomics, and it is in these sorts of eternal discussions that ideas come. It is quite rare that anyone sitting alone thinking in their room comes up with any big ideas. It's really by throwing lots of ideas around. If you have a hundred ideas on the table, then one of them turns out to be really cool.

Gitschier: What other mysteries would you like to address?

Pääbo: What one dreams about is defining the genetic changes that we all share today but that made modern humans so special. That made us colonize the whole place, every little speck of land on the planet, which, after all, archaic humans had not done. They had been around for two million years, but they never crossed the water where they couldn't see land on the other side. Modern humans have been around for a hundred thousand years and we've colonized Easter Island, right?

Gitschier: Not to mention we went to the moon.

Pääbo: Exactly! We're crazy. Nothing really stops us. So there is something really special there in how we behave—to somehow understand that!

Something we also talk a lot about in the group these days is how genetic diversity is structured in humans. I think we are still far too much in the pattern of looking at diversity of different groups and the boundaries between them because of how we have sampled and how we have looked at things. I think, in a way, it is sad that people interested in population history have gone out and sampled according to preconceived ideas of what groups are there, be those linguistic groups or racial groups, and of course if you sample like that you come up with some differences between groups, and say yes, they are there. Rather than going out and just sampling without regard for anything other than geography.

Gitschier: So you mean, just getting a map and sampling a person at every grid point.

Pääbo: Yes, and the logistics of doing that over a whole continent are almost impossible.

But coming back to Egypt again, what I would really like to do is have a boat and sail along the Nile from the Mediterranean, where people are really "European-like" to the source of the Nile in Lake Victoria, where people are really "African-like", and sample every 50 kilometers along this corridor through the Sahara and just see how this transition occurs. Are there sharp borders or is there a gradient? And that would be a powerful thing to look at, and it would be feasible.

Gitschier: Well, if Craig Venter can sail around the world collecting microbial samples, you should be able to get a boat for a trip up the Nile.

Pääbo: I'm not as independently wealthy as he is!

Gitschier: Now, in our final moments, I want to ask what has been your favorite project.

Pääbo: I tend to think the current project is the favorite project. Every project has this manic thing about enormous expectations in it, that often are not borne out to the extent that you imagine, but it's what drives it. And then you come down to the reality of things.

But clearly now, I would say, being able to see the Neanderthal genome is something that just a couple of years ago I wouldn't think would be possible in my lifetime. And now, it is.

Gitschier: What do you think it is about Neanderthals that excites people?

Pääbo: Quite recently, only some 2,000 generations or so ago, there were some other humans with us who were similar, but clearly distinct from us. It gives us some perspective.

Sometimes I like to make the thought experiment—that they made it another 2,000 generations and were here! What consequences would that have? Would racism against Neanderthals have been even worse than the sort of racism we experience today, because they truly were a bit different, or is it that if we had had something like that that was another human form, then perhaps we wouldn't have been able to distance ourselves so much from the great apes as we do now—not making this enormous distinction we do now between what we call humans and all other organisms which we call animals.

It could have gone either way—we can never know, but these are things it is interesting to think about because it puts these issues in our society in perspective. Perhaps somewhere there is the fascination.

The Whole Side of It

An Interview with Neil Risch

Because any two individuals differ in their DNA sequences, study of the variations in sequences within a given population and between populations can provide insights into their origins and relationships. This information can also be coupled with data on the prevalence of disease or physical traits within populations, leading to hypotheses on the effect of DNA variants during human evolution. This interview touches on these questions as well as the controversy over what genetics tells us about the meaning of "race."

Interviewed March 8, 2005

Published July 25, 2005

TRYING TO TRACK DOWN NEIL RISCH IS THE STUFF OF LEGEND. E-mails to him can bounce back with the comment line "overwhelmed by e-mail." Phone calls lead to voicemail, and faxes to a tepid response from his assistant. It's not that he's reclusive or off windsurfing, he's simply a whirlwind of genetics ideas and activity.

Indeed, when the University of California at San Francisco was looking for a director of the new Center for Human Genetics, Neil's name quickly came to the top of the heap. With wide-ranging experience and interests, he was described by one of the field's founding fathers as "*the* statistical geneticist of our time." It didn't hurt that he is a mensch.

I managed to trap Neil in his bright new office on the ninth floor of the west tower off Parnassus Avenue. Still spartan, with only a computer, a phone, and a chair, the office's view spanned Golden Gate Park, the Marin headlands, and the Pacific Ocean. The vista was interrupted only by the jarring copper-clad tower of the new museum under construction in Golden Gate Park. It was a brilliant blue, warm afternoon, and I looked forward to spending some one-on-one time with this man, with his infectious laugh and his intellectual stamina.

Gitschier: Let's start with the broad view. What really interests you?

Risch: My passion, really, is the interplay between population genetics and clinical applications—to see the whole side of it.

When I was in graduate school, I came out of math. Three weeks into my first course in human genetics [as part of a new biomathematics graduate program at the University of California at Los Angeles], I knew that it was what I wanted to do. One, I loved the subject matter. Two, I loved the quantitative aspect of it. Three, I loved the intuitive aspect of it. It was almost like I could predict the next lecture. It was a perfect fit for me. My other passion was population genetics, but there weren't many career opportunities in that field back then. So always there was this latent passion for population genetics without the opportunity to act on it.

The thing that has been exciting for me is that I saw ten years ago where the field was going. Having the sequence of the human genome provides the opportunity to look at the variation in that sequence as well, which leads to the marriage of several areas, but particularly human population genetics with disease studies—genetic epidemiology.

But the fields have never been so intimately related as they are now, and I am just thrilled. I get to marry the two things I love to do. In the old days, the NIH [National Institutes of Health] would never fund a study in population genetics, but now it does because you need to understand human population history and genetics to undertake all these studies of human genetic variation underlying disease susceptibility.

Gitschier: So many people want to collaborate with you. How do you choose what projects to become involved with? Did you initiate most of the projects you work on, or do people come to you?

Risch: In my Stern Award address [presented to the American Society of Human Genetics in 2004], I talked about the population genetics analogy of selection and drift. Some things are planned and have a natural scheme to them, and some things are just random. Some things I have become interested in and I have initiated, and other things have come to me and I've participated in them either because it was appropriate, given the setting where I was working, or I was interested in the project and wanted to have a collegial relationship.

There has been quite a range of the projects I have gotten involved in, especially on the clinical side. I think it's been good for me that I haven't focused on one particular area, like cancer or psychiatry. Some people could say, "He's like a dilettante," but I think it has given me a broad perspective on the clinical application and the commonalities in terms of the issues involved across diseases, and some of the unique aspects. I value the fact that I've been able to be involved in a lot of different things. I feel incredibly fortunate to have become established in a field so that I've had a lot of opportunities in terms of different people approaching me for collaborations. It's been great.

Gitschier: You are an Ashkenazi Jew and belong to a conservative temple here in San Francisco. You also are well known for your work on diseases, such as torsion dystonia, that affect this particular population. Do you think you gravitate to these problems because of your ethnicity?

Risch: Early on, I was interested in population genetics and I knew about this debate about the presence of the lysosomal storage diseases in the Jews. Why did the Jews have all these diseases?

Gitschier: I just assumed it was selective advantage. What was the debate?

Risch: Between selection and drift—as it always is.

This actually started with the dystonia work. At the time, there was a raging question about the mode of inheritance. They knew it [severe early-onset idiopathic torsion dystonia] was more prevalent in Ashkenazi Jews. Some people thought it was dominant, but there was a major paper that said it was recessive just like all other major Ashkenazi diseases.

And I was interested. It was a nice statistical problem, and it was in the Jewish population, which I was interested in both for scientific and historical reasons, and because of my own identity.

So we did a study. Susan Bressman, my colleague at Columbia, systematically went out and clinically examined all the first- and second-degree relatives of all the early-onset Jewish cases—parents, siblings, children, nieces and nephews, half-sibs, uncles and aunts, maybe grandparents. We analyzed the data, and the rates of dystonia in the relatives were the same in all the first-degree relatives; the siblings did not have a higher risk. About 15% for everybody, across the board. We concluded that it was autosomal dominant with low penetrance [30%].

We did a formal segregation analysis. We could clearly, overwhelmingly reject a recessive model. And at the time I thought, "Well, this is fascinating!" because we had a dominant disease with a founder effect in the Jewish population. And we even suggested that this would be valuable for gene mapping—by linkage disequilibrium [LD] analysis. And after mapping the disease, we found strong LD right away. So this was very clear evidence that we were dealing with a relatively recent founder mutation.

I know this is a long story, but it was through that project that I got interested in Jewish population genetics.

Gitschier: It seems as though every time I open the science section of the *New York Times*, you are featured in it. These articles, at least lately, focus on your adherence to an often politically incorrect idea, such as the genetic basis for race or the way NIH should spend its money on diseases of addiction. Do you deliberately choose controversy?

Risch: I think historically I have avoided it. Perhaps this is what job security offers you—the opportunity to get involved in potentially more controversial questions. And I think I've decided that playing it safe is not the way to go. I just don't believe that anymore. These are big important subjects and I just don't think they should be avoided.

Gitschier: Let's talk about the former, the genetic basis of race. As you know, I went to a session for the press at the ASHG [American Society for Human Genetics] meeting in Toronto, and the first words out of the mouth of the first speaker were "Genome variation research does not support the existence of human races."

Risch: What is your definition of races? If you define it a certain way, maybe that's a valid statement. There is obviously still disagreement.

Gitschier: But how can there still be disagreement?

Risch: Scientists always disagree! A lot of the problem is terminology. I'm not even sure what race means, people use it in many different ways.

In our own studies, to avoid coming up with our own definition of race, we tend to use the definition others have employed, for example, the US census definition of race. There is also the concept of the major geographical structuring that exists in human populations—continental divisions—which has led to genetic differentiation. But if you expect absolute precision in any of these definitions, you can undermine any definitional system. Any category you come up with is going to be imperfect, but that doesn't preclude you from using it or the fact that it has utility.

We talk about the prejudicial aspect of this. If you demand that kind of accuracy, then one could make the same arguments about sex and age!

You'll like this. In a recent study, when we looked at the correlation between genetic structure [based on microsatellite markers] versus self-description, we found 99.9% concordance between the two. We actually had a higher discordance rate between self-reported sex and markers on the X chromosome! So you could argue that sex is also a problematic category. And there are differences between sex and gender; self-identification may not be correlated with biology perfectly. And there is sexism. And you can talk about age the same way. A person's chronological age does not correspond perfectly with his biological age for a variety of reasons, both inherited and non-inherited. Perhaps just using someone's actual birth year is not a very good way of measuring age. Does that mean we should throw it out? No. Also, there is ageism—prejudice related to age in our society. A lot of these arguments, which have a political or social aspect to them, can be made about all categories, not just the race/ethnicity one.

Gitschier: I have heard you say, "Don't politicize the human genome."

Risch: I have a strong problem with the way politicians use this information. [Former President] Clinton, for example, when the first draft of the human genome sequence came out, made a statement about how all people in the world, in terms of their genetic makeup, are 99.9% the same. His intent—to reduce conflict among peoples—is noble. People on the left, anthropologists and sociologists, do the same thing. They use the 99.9% figure as an argument for social equality. But the truth is that people do differ by that remaining 0.1% and that people do cluster according to their ancestry. The problem is that others could use that information to create division.

Gitschier: Do you ever feel that the press misrepresents you?

Risch: Don't we all feel that to some extent? They always take the simple side, which leads to misinterpretation. So there are risks in talking to the press, especially on controversial subjects.

Gitschier: You have a brother who is an academic, and at one point you and he were both on the faculty at Yale. Is that genetic? Tell us about the environment that you grew up in that ultimately led to producing two academicians. I don't know what your brother's field is.

Risch: So interesting. You tell me, is this "nature" or "nurture"?

My brother, my only sibling [Harvey Risch], is 20 months older than I. Mathematics was his skill set also. He went to Cal Tech, starting a year before me—both of us math majors. Then he decided to go to medical school, so he did additional coursework, and we graduated simultaneously. He went to UCSD to medical school, and had to do a thesis—so he came to UCLA in my department and lived with me. Then he applied to do a PhD in biomathematics in my department [but went to the University of Chicago instead]. His first paper and my first paper appeared in the same issue of *Annals of Human Genetics*, and we didn't even know it. Then Harvey decided to do epidemiology as a post-doc, while I was learning epidemiology at Columbia. Our grades, SAT scores, GRE scores—everything pretty much the same.

Gitschier: Sounds like the premise for a simple quantitative analysis. Do you think you are both hardwired to do mathematical problems? Or did your family just sit around at dinner doing math problems?

Risch: Not at all. My mom [Sonia Risch] was very artistic and intuitive. Great artist, writer, actress—and brilliant at all of it. My father was a clinical psychiatrist. But if you look at my family history, on my mother's side there are a lot of MDs and on my father's side there is a lot of math. So if you want to make a genetic hypothesis here, my brother is the confluence of both, maybe, and maybe me, too. My brother and I would talk about stuff a lot. My mother would say, "Oh, they're talking Fortran."

Gitschier: You have been working with the epidemiologists at Kaiser Permanente [a health-care provider] in Oakland for the past seven years. I understand you spend every Wednesday there. How did this collaboration come about, and what is it you are trying to accomplish? Other than numbers, is there something the Kaiser resource can do that say, Iceland, couldn't?

Risch: Can it ever!

Gitschier: Why don't you tell me what's so cool about it?

Risch: When I went to Stanford, at the back of my mind was this issue about Kaiser. This comes from my epidemiology background. This is the advantage of being multidisciplinary. If I could push anything, it's the value of seeing the links between various disciplines and marrying them. One, Kaiser's membership is a cohort—you don't have to construct it from scratch. And it's followed over time, for many people over 20 years. Two, they have computerized databases where every contact a patient has with the health-care system, every inpatient, outpatient, and pharmacy visit, every visit with doctors outside the Kaiser system gets recorded, every X ray, every lab test, basic biochemistries—all computerized. Three, it is the health-care provider for one-third of the Bay Area and a very good representation of Bay Area population. It is missing only the very high and very low end of the socioeconomic ladder. All major ethnic groups are represented. To me, it's the most wonderful laboratory for doing population genetic and genetic epidemiologic research.

As you can tell, I'm a very strong believer of inclusion of a variety of people of varying racial/ethnic backgrounds in research. There is everything to be gained from doing so—not just politically, but scientifically. Also, ethically it's the right thing to do. And I'm concerned in this whole discussion that people may be scared off from the genetics research and that's the battle.

One big issue that I think will go a long way towards addressing this problem is to do everything we can to recruit more minority scientists to human genetics. For the research to have credibility in minority communities, there must be representation from those communities among scientists. And I want to be involved in that process also.

Gitschier: In Iceland, you've got all the ancestry data, and so you can do traditional linkage analysis, but in Kaiser, you're going to do association studies.

Risch: That's right. Because of the way the technology is moving, this is a tremendous resource for doing that.

Gitschier: This is a very long-term study. There must be an element of "Oh, I see this gelling" and it just can't go fast enough.

Risch: You're right, it's been a long process. There are complexities and financial issues.

This is really the best opportunity we have in the United States to do something along these lines, and it's been a little frustrating that it's been difficult to get the support and funding for it, but I'm patient. Because when you believe in something and know it's right, you have the patience to see it though.

Ready for Her Close-Up

An Interview with Elaine Strass

Most scientists belong to professional societies that foster interaction and support publication of their research findings. In the United States, two societies serve as hubs for the world of genetics: The American Society of Human Genetics, which tethers such diverse interests as medical genetics, human molecular genetics, and genetic counseling, and The Genetics Society of America, whose focus is primarily on so-called "model" organisms, such as fruit flies (Drosophila melanogaster), zebrafish (Danio rerio), worms (C. elegans), and bacteria (e.g., Escherichia coli). This interview looks at the workings behind the scenes for these two vital organizations.

Interviewed October 24, 2007

Published February 29, 2008

WHEN I INVITED ELAINE STRASS FOR AN INTERVIEW, I had no idea she was planning to retire in the coming year. Elaine has been the hidden force behind both the Genetics Society of America (GSA) and the American Society of Human Genetics (ASHG) for almost 20 years, by serving as Executive Director for both organizations. Fresh, articulate, and cheerful, Elaine has a lightning wit, great people skills, and zest for her work. She is our societies' strongest champion, yet many of you may be unfamiliar with her.

To catch up with Elaine, I flew to the ASHG meeting in San Diego in late October with a bit of trepidation and a lot of sadness for the residents there, as fires were devastating the outlying areas. From the window seat on my evening flight, I easily spotted at least half a dozen blazes in the distance. It was a spooky sight, reminding me of the strong natural forces that shape our planet, but I couldn't help likening the vista to red fluorescent probes beaded along interphase chromosomes against the black landscape of a FISH experiment.

Unannounced, I located Elaine the following morning at the ASHG headquarters in the convention center. We popped a few batteries in the recorder, turned on the machine, and we were off—literally! At a quick pace and with effervescent description of the success of the meeting so far, she swept me along to show me the new booth that ASHG had designed to advertise the next International Congress of Human Genetics. She was ready for her close-up.

Gitschier: How long have you been the executive director?

Strass: I became the executive director of ASHG and the Genetics Society of America in 1992, when Gerry Gurvich, who had started the Washington office in 1983 for the two societies, decided to retire. It was the first time either society had had an office, and they decided to do it together for economies of scale. They shared staff and that is still true to this day.

Gitschier: How did you land this job?

Strass: Well, I started off as a concert pianist, and I wasn't very good and I needed to get a day job. The only thing I really knew how to do was type. So I went back to school and learned a little bit about computers and word processing, which was *the* thing then. I was taught on a Wang system.

I got various jobs doing word processing. I was staying home with my kids, so I wasn't working full time, but I was having a wonderful time, and I was also doing a lot of concertizing in the community.

Gitschier: In Washington?

Strass: Yes, but I was never very famous. You have to put this in perspective, Jane. I'm a very good pianist, but I'm not a *great* pianist.

Gitschier: What kind of concertizing did you do?

Strass: I was the official accompanist for the State of Maryland for competitions they had, like violin concertos, cello concertos.

Gitschier: Had you been a conservatory student?

Strass: I was a graduate of the University of Illinois. Bachelor of Music in Performance Piano. I *loved* playing the piano. I did accompanying stuff, pianist for shows. It was a lot of fun for me.

Anyway, to get on with my brilliant career.... When my third child was in second grade, I went to work part-time for a law firm in Rockville.

Then, in 1981, I heard about a job that I really wanted. It was for the Society for Neuroscience, and it had to do with computers, sessioning for abstracts for their annual meeting. They were panicked because they had 5,800 abstracts, and they had never seen so many abstracts in their lives! This was the first year that they were going to be using a computer. They hired me.

That is also where I met Gerry Gervich. We worked together for about three years. And later Gerry met some geneticists—including Art Chovnick, who was the most instrumental in putting together that first [genetics] office. They got the idea that ASHG and GSA would chip in and have employees and then everything would be official—the registration, documentation, computers for membership. Gerry had this all mapped out in her mind. She wanted to hire me then, but she didn't have the budget.

Then Gerry called me one day and said, "I've got the budget, I'm going to hire you." So in 1988 I started working for ASHG, but not yet for GSA—raising funds, doing committee work, and supplementing what she had already set up. I loved my job; I was so happy. I loved working with the geneticists and I became fascinated with genetics, even more so than I had been fascinated with brain science.

After several years, Gerry decided to retire. She asked me if I would take the job. I didn't think I had the right stuff. It's a really tough job. Gerry had a lot of faith in my ability, and both boards [of GSA and ASHG] agreed that they wanted me. The first meeting I did was the San Francisco meeting of ASHG.

Gitschier: Do you go to the meetings?

Strass: I go to all the meetings. I go to the yeast meetings. I go to the *C. elegans* meetings, which we've started to do. They meet every other year. They usually meet at UCLA. As long as we can do it this way, we will, but some of these meetings are threatening to grow even more, so we may not be able to do the campus meeting.

The *Drosophila* community has a meeting once a year and they have grown to such a size—there are 1,500 people now—that they meet partly in a convention center, partly in a hotel.

But some of the other meetings are less sizable, such as the yeast meeting, which has between 800 and 900 people every other year, and what they prefer to do is to meet on a college campus in the summer and to stay in the dorms. The cost is so reasonable for the students who go. It's a good deal.

GSA handles all these different kinds of meetings. There's also fungal genetics, about 800 people and they meet at Asilomar every year, but they may have to change their venue because of their growth.

Gitschier: What about the zebrafish community?

Strass: Zebrafish is going to contract with GSA and have their meeting organized by GSA in two more years.

The campus meetings are great, but the problem is that when you start to have too many people coming, you really need to be in a tax-free situation where you can receive money and not pay income tax on the receipts from the registration fees. So you need a tax-exempt carrier for the money. And that's why GSA became so important in that community, with all those little organismic meetings, which are extremely critical to the development of those fields.

Gitschier: I assume GSA has their own president and board of directors?

Strass: When Gerry Gurvich put them [GSA and ASHG] all together in 1973, she modeled them very similarly. The idea was to keep them totally separate, because they do not have very many overlapping members, and they have separate goals.

The governance structures are pretty much the same. The main difference is that GSA, in their election process, has two people running for president, but ASHG has only one. Of course that is always being discussed by the ASHG board: Is that the best way to represent the society? There are ways to look at it from both sides.

The nominating committee is wonderful—ASHG and GSA both have these committees and they take their job very seriously. We always have very good leaders who are dedicated to the societies and their missions. It is always a wonderful thing to see this all unfold with their elections.

The argument for having one candidate only is that [with a two-person race] you can work your way through a community generating non-winners, and sometimes there are hard feelings. We really don't want to do that.

People work themselves up through the ranks of ASHG, we don't ever nominate a president who isn't familiar with the work of ASHG, who hasn't been to the meetings and participated in the work of the society. So we are very familiar with a candidate and truly recommend someone who is just great. And we keep lists from year to year, so sometimes, when a name gets on, it might take five years until they work their way to the top, so they really have been elected in a sense. It's a very thoughtful process and I recommend keeping that. I think if people knew the process, they would agree that it is a good one.

Gitschier: What is the membership of GSA?

Strass: Right now 5,100.

Gitschier: And ASHG?

Strass: 7,200.

Gitschier: Tell me exactly what you do?

Strass: My job is very exciting. I run the office in Bethesda. We have 14 employees. We are a very well-oiled machine. Our IT department is our biggest department. Everything is done over the Web—dues, memberships, meetings.

One of the most important parts of running a non-profit organization is to document all financial transactions. And then there is the paying of bills for the meetings. ASHG meeting here in San Diego, for example costs $2M. We don't really make much money from the meetings.

I make sure things happen. I listen to what the board wants. I give them my ideas. Sometimes my ideas are very well received. Sometimes I have to re-introduce an idea several times.

People come to me and I serve as a funnel, and I see that as a very important part of my job. Also, when I read something in a magazine or newspaper, or if I get an idea from the Internet, as the Executive Director, I have the authority to institute a lot of ideas.

Right now it's a wonderful time for us to be doing this because of the Internet. It has changed the way societies do business and has made everything much cheaper to run and extremely efficient. We're delighted.

We never take Fedex submissions of abstracts—we used to have 2,000 Fedex envelopes all arriving on the same day at the office! The fascinating part was the fear we faced—oh, we'll never be able to do this electronically! What if the disk breaks, or something? It's in the ether, it's not concrete. But we made that transition very, very quickly.

Gitschier: I assume you like your job.

Strass: I love it!

Gitschier: Why?

Strass: First of all, working for geneticists is for me a big thrill. Don't forget, when I was a housewife sitting in my garage watching people drive up and down my street, I didn't see many Nobel Prize winners!

But I do now!

And one of the biggest thrills at GSA this year has been the realization that GSA has been making awards, and the recipients then go on to win the Nobel Prize. The GSA Medal went to Bob Horvitz and then he won the Nobel, and then to Andy Fire, and he won the Nobel. John Sulston was another awardee, and then he won the Nobel. This year we awarded Oliver Smithies, and he just won the Nobel! We beat the Nobel to it! It made us feel terrific.

The GSA has three awards traditionally. There was the Thomas Hunt Morgan Award, which is a medal and it's been given since the early '60s, and there is a picture of a fly on the medal and on the other side is Thomas Hunt Morgan's portrait—4.5 inches in diameter, made out of pewter. And this is the big one. You get this award when you have shown lifetime contributions to the field of genetic research.

The GSA Medal was developed a few years later. Why should we have to wait for a whole lifetime for contributions to occur before we can honor someone? The idea for the GSA Medal was a breakthrough within the last 15 years. And that was the one that Horvitz won, for apoptosis, programmed cell death. He had other things, too. It was not difficult to give Bob that award!

And Oliver Smithies, of course, for his great contributions—he got the Thomas Hunt Morgan Medal for lifetime contributions.

Gitschier: So the GSA Medal is like the Curt Stern Award that ASHG gives.

Strass: Yes, it's the parallel to that.

GSA wanted to honor people, starting about six years ago, for service to the community. It's called the George W. Beadle award. We have people who have made enormous contributions, who have made the lives of the scientists so much easier and even possible. Like, if you ran a stock center for 20 years, you might not get a scientific award, but, my goodness, your donation to the community is so enormous! So that's why the Beadle Award was developed and every year we come up with some really great winners.

Gitschier: With GSA having all these separate meetings, how do they coalesce to make decisions about things?

Strass: It's a good question. Let's take a step back and look at the bigger picture. We have different models for societies.

We have Society for Neuroscience with 48,000 members. When they have a meeting and 34,000 people come, the city knows that they are there because they have taken every single hotel room. They have so many poster presentations that they have a complete poster session in the morning, then they take it down and another complete one goes up in the afternoon, and that goes on for five days. The scale is different from both ASHG and GSA.

There is a drawback there because when you go to meeting of that size, it's hard to run into people. They have specialties, too, but they elected to keep everything together. I remember when I worked there, there was always the threat that the behavior neuroscientists would split off! "We don't like the way the board is treating us and we're going to split off!" But they never split off.

Still you know where you fit in, your little corner in the very large meeting.

ASHG is kind of in the middle. We get together once a year, we know we can bump into people we want to bump into. We have little cubbyholes to leave messages.

However, we also have American Society for Gene Therapy, NSGC (National Society of Genetic Counselors), HUGO (Human Genome Organization), the American College of Medical Genetics—all these other groups who are very close to ASHG. But the ASHG meeting is *the* research meeting, and we have chosen not to do the neuroscience route, which is to include everyone at one meeting.

The way I interpret it—genetics is nature's way of making diversity, and that's what we've got in the genetics community.

All these organism guys, they really don't want to meet with each other. The yeast guys don't want to meet with the worm guys and they don't want to be with the fungal guys. They call *Clamydamonas* guys the "pond scum" guys. The Drosophilists have their own wonderful community. They have their own board of directors, even though they are not incorporated. It's very loose, they don't have to file reports.

So the question is—could there be a really large model organism meeting? It would be about 9,000 people, and I've given this a lot of thought. Possibly serial overlapping meetings. But nobody is hot to do that. They like their small meetings, which are very predictable, good ways for students to present their first poster or talk, and to teach students how it is done—how to meet the right people and publish.

Gitschier: So, with all these independent meetings, how does GSA come together?

Strass: There is the journal *Genetics*. GSA was started in 1931 and the journal in 1916! And all of those old issues are on the Journal Web site through Highwire Press. So the Journal acts like a coalescing factor. And the GSA board has been discussing this for years, and that why they came up with the Model Organism to Human Biology (MOHB) meeting, which meets in San Diego every other January.

This was actually a decision made while Mark Johnston, a yeast researcher, was president of the GSA, and he happens to be a very visionary person. He, along with many others on the board, the Drosophilists, thought it was time to bring things together. Part of it had to do with the fact that everything is getting sequenced, and this made a huge difference in the way we look at genetics research and the future as geneticists and genomicists. There's a lot more that we have in common because of these important features. And the Journals acknowledge this, but having a meeting where you have the human and the Drospholist and all talking about the same gene—in yeast and worm.

At the last MOHB meeting, I was floored.

Gitschier: So you went to the sessions?

Strass: I always go to the sessions. I don't know how I can understand them, because I'm not trained as a geneticist. To me, genetics is the mechanics of the tinkertoys of life. It's a huge puzzle and I *love* the idea that it could be decoded, that it could be sequenced. The idea that somebody would invent PCR. I understand enough to see the wonder of all this.

Gitschier: You're tearing up!

Strass: I'm just an old sentimental fool.

To see the changes in the field! I'm almost an outsider. I don't understand enough to really appreciate everything that's going on, but when somebody gets the Nobel Prize, I understand.

Gitschier: Why are you retiring?

Strass: Well, I'm more aged than I look! I've got old arteries. I have to take care of myself. This is a big job. A $9M budget—two budgets! And a lot of travel, if you go to all the meetings. I go to everything—I wouldn't miss anything—it's too exciting! And that's a problem I have; if I could step back and not be as involved, it would be better for me. I can't seem to do that! So, I announced two years ago that I was going to retire.

There are other things I want to do. Go back to music and work in the community. I'm starting a business called "Dial-a-Daughter." I will take old people who can't drive any more who want to do things with me, like go to the opera or out to lunch. I will devote my time to helping these people *live*.

This will be my new life, and I of course will continue to be a member of GSA and ASHG.

There are so many people who have helped me—these people are wonderful. I adore geneticists! Warm people who just helped me understand anything I didn't understand and helped me get through the things I wanted to get through. This has been a real "love in," for all these years.

For me, it's the end of an era, and when I look back on my life, this has been the best part of it. There is no question that what I have here is rewarding and exciting—it's a dream job, and it's all because of these two societies.

Knight in Common Armor

An Interview with Sir John Sulston

Forty years ago, Sydney Brenner wanted to understand how genes orchestrate the development of an organism at the cellular level. To achieve that goal, he advanced a new genetic model, the tiny worm C. elegans, *which is virtually transparent during development and consists of only about 1000 cells at maturity. This interview burrows into one aspect of the* C. elegans *work during that fertile period, the painstaking effort that led to a complete description of cell lineages during the worm's development and supplied a generation of geneticists with questions to answer.*

Interviewed July 11, 2006

Published December 29, 2006

I SPENT TWO OF MY THREE MONTHS on sabbatical in Cambridge gathering my courage to invite Sir John Sulston for an interview. How do you approach a man who has spearheaded and labored with his own hands on three major genetics projects, won a Nobel Prize, been knighted, and even had a building named after him? Perhaps it was the salutation that gave me pause, before settling finally on "dear sir john" in an e-mail.

Capped with a head of vigorous white hair and a face framed with a matching beard, Sir John has a rock star recognizability. Thus he captured my attention one Saturday afternoon as my 12-year-old daughter and I bicycled past the Fitzwilliam Museum, and he pedaled by in the opposite direction. "There is the man I'm hoping to interview!" I said as I pointed out the man in a red shirt.

I told her he was Sir John Sulston. "Is he a prince?" No, just a knight.

That he had won a Nobel Prize. "Is he rich?" I don't think so.

And that he had won it for his work on worms. "Does he know that worms are segmented?" Ah, a tricky question! Not his worms, tiny creatures called *C. elegans*. I started to wonder if maybe she should do the interview.

Sir John was catapulted into the public light as the spokesperson for the human genome in the UK. His experiences in defending the public genome efforts against the assault of privatization and patenting, chronicled in his book *The Common Thread: A Story of Science, Politics, Ethics and the Human Genome* (Black Swan, 2003) were transformational. Now officially retired from the Wellcome Trust Sanger Institute (WTSI), he is absorbed in policy-making for the UK government, the World Health Organization, and a variety of non-governmental organizations (NGOs). An activist and humanitarian who still rides his bicycle to work, who is content to share a small office, and who ponders whether our species will survive the coming century, Sir John is an inspirational man, one who leads by example.

Gitschier: I'd like to start with the process of looking at worms and tracing their lineages. When did this work begin?

Sulston: It was mid-'70s, I think. I'm a bit vague because I did all sorts of different things after joining Sydney's [Brenner] group in 1969. One of those projects led me to being interested in the cell lineages, because I wanted to know where the dopamine-containing cells came from.

I was in the midst of a hobby project, a method for displaying catecholamines as bright fluorescent adducts. After some fiddling around, it worked and gave some beautiful patterns. It was clear that a very small subset of neurons contained dopamine and some contained serotonin. That was really no big deal.

However, what *was* potentially a big deal was that some of those cells appeared only after the embryo hatched. There had been a general view floating around, one of those urban myths, that there really wasn't any neural development after hatching. So I thought we should follow this up.

I started looking at these cells with Nomarski microscopes, of which there were several around the lab. I was lucky enough to stumble on a way to just look at them, without any fancy photography, and discover where the cells came from.

But then, we realized that other sections of the nervous system also developed post-hatching. The long and short of it was that I found myself able to watch dividing cells.

Gitschier: At some point you moved from looking not just at neural cell divisions to looking at the whole thing.

Sulston: That's right. Just bit by bit. The very first lineaging was the ventral cord, a set of motor neurons that make the organism move in a sinusoidal wave, backwards and forwards.

Just after that, Bob Horvitz showed up. He was a hard-core molecular biologist. He thought all this zoology stuff was a bit of rubbish. He was fairly amazed that a guy was sitting there studying cell lineages.

It was quite an interesting rapprochement. Bob wrote it up as a nice thing for *Genetics* [1980] in which he described how this happened, how at first he was completely bemused, how if it didn't involve a scintillation counter it wasn't real science. He hadn't realized that this was really precise, really digital information coming out. Not sloppy stuff.

So *he* then became the driving force to push it to the next stage and do *all* of the post-embryonic lineaging. He did some, I did some.

Gitschier: You talk of this period of a year and a half when you focused on the embryonic development, to complete the entire lineage.

Sulston: That was later. This is going back to the history and why people hadn't made progress with the fancy Nomarski microscopes and why I did. Because of my initial interests, I was looking at the larvae, which others had completely ignored.

Gitschier: Because that stage was thought to be too late?

Sulston: Yes, because there was this view that the nervous system wasn't developing there anyway. Also, it was difficult because the worms are wriggling around. People were interested, but they killed them or anesthetized them, and of course, nothing happened because the worm is unhappy, or dead, or about to die. And the result was it was all fairly hopeless.

What I hit on was the idea of putting living worms down on an agar pad with some bacteria for them to eat, so they were happy! And although they were still moving, they weren't moving so fast that you couldn't follow the cells.

And this is what Bob Horvitz found me doing, just after I had completed the ventral cord work, and he joined in. And then we passed this technique—it seems a rather grand word for just looking at worms on an agar patch—on to people all over the world and it allowed them to start looking at post-embryonic development and make use of mutations and so on.

So there was great excitement about this. It was the little wonder of that particular year that suddenly we could see the cells. And it linked the work John White was doing on reconstructions of the neuroanatomy. He could, from the output of the lineage work, say which cell was which and correlate the position of the twig on the lineage tree and the type of motor neuron it was producing.

Gitschier: And these reconstructions were by EM [electron microscopy]?

Sulston: Yes, and there was a guy named Nichol Thompson on whom everything hung because he was able to cut very long series of sections without losing *any*. It was a great deal of work, but it meant that people could go through those sections, make photographs, trace through the stack. Because there were no gaps, you could actually follow through the profiles of the axons.

So, then we wanted to know where these neuroblasts came from—to go back into the embryo. And this had been where people had thought to start because it was easy to have eggs survive in their hard egg case under the microscope. But the problem was that it was hard to follow the cell lineage in a ball of cells with really no phenotypic characters.

But there came pressure from the community to move forward on the embryo, because of a dispute. Someone had claimed to have the lineage of the intestine, and people were doubtful as to whether he got it right, so I was asked to arbitrate. So I ended up doing my first bit of embryonic lineaging using exactly the same method, just putting the egg under Nomarski optics and watching and drawing.

I can show you the patterns, if you want to see the way they're recorded.

Gitschier: Yes, love to!

[John opens his small Sony PC to a record dated 5 June 1980. There are a series of colorful small drawings with circles and arrows, reminding me of an American football playbook. We delve into the drawing (Box 1).]

A Page from the Notebook of Sir John Sulston

Sulston: Let me explain what you see on this page. This is a particular bunch of cells at a particular time. Here are the minutes during which I am watching it. Drawn every five minutes or so, over a span of about three hours. I broke it into stages—I followed a particular cell and its descendants.

And then I put these different bits together. It was based on the assumption that the lineage would be fixed. And, of course, as I did it, I checked to make sure that things really did always happen the same way, and indeed they did, apart from very rare cases where cells determine their fate by their interaction.

Gitschier: So you've got red pens, green, blue, black pens, drawing quickly every five minutes. This is the kind of mundane thing I'm interested in.

Sulston: I thought it might amuse you. You've got red, green, black, blue, and purple on this one.

Gitschier: This looks like Russian to me [as I pointed to a cell labeled "MStpaaapa"].

Sulston: These are the names of cells. MSt is one of the primary blast cells, and that means the posterior daughter, and the anterior daughter, and the anterior daughter and the anterior daughter and the posterior daughter and the anterior daughter. And there is a question mark! Wasn't sure if I got that right or not.

I should explain the colors. They are depth. Red is out the top, then down a bit is green, down a bit more black. This is the problem with movies, you get all these images at different depths and then you've got to try putting them together. Now, the drawing you're putting together over a limited area at the time. So I'm quite certain of what I'm looking at. However, I do have to follow about ten cells at a time.

Gitschier: What are the hatch marks here?

Sulston: That is a cell death, one of the things that the worm has become rather well-known for. Programmed cell death was known about in mammals, but there were no handles on it. But because the worm has a fixed cell lineage and we knew which cells were going to die, we could look for mutants that change that pattern. The hatching means that the cell has gone very refractive—very bright—in Nomarski image.

Gitschier: Are these the same color codes you used with the larvae?

Sulston: Yes.

Gitschier: So you are actually thinking in color here.

Sulston: It's an approximately spectral range—red at one end, violet at another—so it is easier to remember. A cell higher in the focal plane is red.

These pictures have actually been quite interesting. In fact there is a whole bunch of these notebook pages in Kettle's Yard in Cambridge ["Lines of Enquiry: Thinking through drawing," a fascinating exhibition by Barry Phipps http://www.kettlesyard.co.uk/exhibitions/archive/linesofenq.html].

Gitschier: How did you settle on this kind of representation?

Sulston: Desperation.

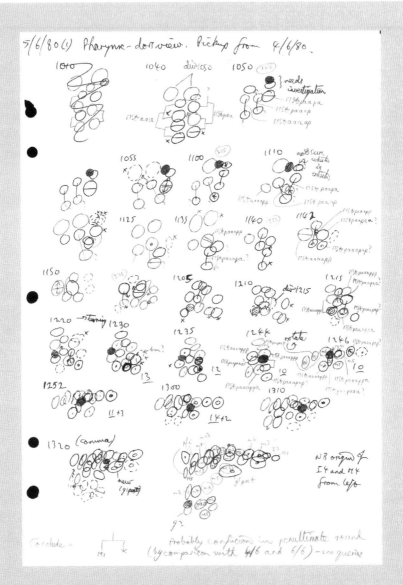

The larva is more or less flat, and when a cell goes past another, you can keep on drawing. But what happens when a cell goes over another cell, oh dear, I get lost. So, I grabbed the nearest color pen and said, "Right, if it goes up it's red." When I came to the embryo I was really equipped to code it. So this is a ball within a ball—ten cells sitting with several hundred cells at this stage.

Gitschier: Love the squiggle—what is that all about?

> *Sulston:* I made a mistake. And I started again.
>
> *Gitschier:* I see you are picking up from where you left off on the fourth of June. You had to find a new embryo and maybe wait around for it to be of the right age.
>
> *Sulston:* I realized that I could come straight in if I picked the right stage embryo. And there is quite a business of doing this, sorting embryos and looking around for the one you want and finding the right orientation. You can pick up landmarks that are unique.
>
> *Gitschier:* So you started your day picking the one or two embryos that you were going to look at.
>
> *Sulston:* That's right. Now on this one, I want to know what this particular cell is, and that is all I want to know. There is something called M1 coming out, sister to a cell death. There was something about that I didn't know, but here I can make a conclusion.

Gitschier: Now would this page have been in a bound laboratory notebook?

Sulston: These are all loose leaf. If you had come to the house I could have pulled these things out. All my archives are there at the moment. They'll land up at the Wellcome Trust after a while. So if the house catches fire we'll lose them all. It's a bit doubtful how many of these you really want.

Gitschier: Well, how many are there?

Sulston: Hundreds. This is really only of historical interest.

The technology for studying the embryo has improved. John White built a 4-D microscope, scanning up and down the whole embryo. And with nice software, you can ask for images in any series you want. And this really put embryonic lineage following on better footing. This came out about ten years after my work. Now there is yet another step, just happening in Bob Waterston's lab in Seattle. They are completely automating lineage mapping by green fluorescent protein-tagged histones.

Gitschier: Did your mind ever wander while you were doing this?

Sulston: You have to be quite focused. I did really button it down. You don't need to draw many pictures when the nucleus is just changing its morphology, but you do have to be really on the ball when the cell goes through its anaphase and telophase because that is what you must follow in order not to be confused.

Gitschier: Were you in a room by yourself?

Sulston: Oh, yeah.

Gitschier: For all these kinds of experiments?

Sulston: I always was alone. I find it impossible if there is distraction.

I remember there was a little bit of conflict about this. Some of the others rather wanted a microscope room, and that is good because it's air-conditioned. In order to keep watching these you must keep the temperature down to about 20°C. It's very easy to overheat [the worms] when you are shining bright light down and the slides get hot. Above 25°C they start to get very unhappy.

And in a lot of labs like Bob Horvitz's lab at MIT, there is a whole row of microscopes, and that is true of anyone now who does this kind of thing. But I noticed, when I visited there, that they really don't talk.

But I remember at LMB [Laboratory of Molecular Biology, Cambridge], people would come in and say, "Oh, look what's happening," and they really wanted to talk about it, and I'd completely lose it! So I said, "Look, I have to have my own room for this." People were feeling, "Oh, the poor guy is a bit fussy," but I got my way, and otherwise I wouldn't have been able to do it.

Gitschier: When was it that you got your own room?

Sulston: When Bob and I seriously started to look at post-embryonic lineage. I think I just made myself sufficiently obnoxious.

Gitschier: So you didn't listen to music or anything when you did this?

Sulston: One thing I did do with the embryonic lineage, I solved the Rubik's cube from scratch. I also made some Archimedean solids, the regular polyhedra. I made them out of thin card and glued them together. So I did find myself little distractions, but obviously things that didn't engage me.

Gitschier: Where was Bob?

Sulston: He was in the room next door. He was doing various other things as well, genetics, while I was doing just solid lineaging. We worked on the first mutants that affected the cell lineage. It was the beginning, I think quite a nice catalog.

Bob is a very thorough man. I remember he made an enormous list of about 150 psychoactive drugs to try on the worm. And any time something happened we would work on it together. I did more of the following up, so in addition to the wild-type notebooks, there are notebooks full of the mutant images.

Gitschier: What do you think it is about *you* that made you particularly good at this lineage work? Not everyone would have the patience and the focus to go through this.

Sulston: Well, I don't think I had anything else much to do at the time.

All of my scientific projects have started as hobbies, something to do in the evenings maybe, and then at some point you find you are taking it seriously and it becomes your day job. It was this business of constantly trying things out and seeing what you *can* do.

Why me? I don't know. I enjoyed looking at the Nomarski objects. Actually, I think they are awfully beautiful. I still find it amazing that you have this optical trick that allows you to visualize what is only a transparent image, after all. Because of the cross-polarized beams, you image the refractive index. The detail you see! I could watch it forever.

And then there was the group thing. I emphasize Bob, but the whole group was involved. And there was this feeling that the lineage was something that people really wanted.

That's the other thing—I actually quite like to please.

Gitschier: You never entered the professorial track, but remained a staff scientist at the LMB. Did that suit your temperament?

Sulston: Absolutely. It was a very good fit for me. If there was something that I thought was important and worth doing, I could just focus on that, and the only pressure was from family life. It meant you had time to sort all this out and not feel the pressure that you had to cut corners or guess.

The negative is that we don't do much teaching. I feel in a way we're not such good citizens because teaching is important. Also, we didn't have independence. As long as I was at the LMB working on worms, Sydney was my boss.

Gitschier: Sitting there in a chair, focusing on this process—it has the feeling to me of a meditative practice. Did you have a kind of spiritual experience?

Sulston: I would only use the word spiritual in the loosest sense. I use the word beauty straightaway, and I think that touches on it. Enjoying this sort of thing is certainly part of humanity. And I'm the secular humanist kind of person, who thinks that we have a love of beauty built into us.

I feel the same way, hiking in Scotland in Ben More. [He points out his laptop screensaver with his photo of the vista, stippled with icons].

Gitschier: And how marvelous to watch something that nobody has ever seen before.

Sulston: Oh yes! The most exciting thing to do is to go somewhere where nobody has been before. It was *hugely* exciting, looking at those cells dividing for the first time and knowing that I could see, I could find out. It's true all the way through science.

Actually you hold it to yourself [speaking in a hushed voice, as he folds his arms across his chest] when you've got something nice, but then within a few minutes you've got to rush out.

Gitschier: What were the times you rushed out?

Sulston: Seeing those first cells divide! Because the previous work looking at fixed cells just hadn't gone anywhere. No one had managed to see anything. But I was just sitting there and suddenly I knew it was all open. Because I could see that first cell divide and I knew I could see its daughter divide.

Gitschier: Did you wait for its daughter to divide before rushing out of the room?

Sulston: Of course! The very first viewing I carried on right to the end. I was entranced. And those divisions are quite quick. Within an hour, it's already beginning to swell up to the next stage.

Gitschier: But another moment might have been when you saw the first cells dying off.

Sulston: That was a rather slower dawning of what was going on. At first, I didn't know what to make of it at all. There must have been some discussion with others about what on earth was happening. I don't know if I or someone else first realized that this might have been a cell death. I just knew that something strange was going on, and this thing had disappeared, so obviously it was a cell death. But I probably wasn't sufficiently well-read to know the background to this and to know about cell death being a programmatic feature in development.

Gitschier: Let's tear ourselves away and move forward to the nematode genome work.

Sulston: At the end of the embryonic lineage work, I didn't know what to do. I wrote that up and it was done. And everybody said, "Well, now of course you do mutants."

But it was a bit competitive by then. There were all sorts of people out there doing it, including my ex-student. I wasn't sure I really wanted to go and compete. It's a bit of a mingy thing to say, but there you go.

I like to do things on my own, you see. I don't like being in the middle of buffaloes on the plain kicking up the dust. So I was looking around for something else to do more on my own.

A rather dramatic flash came when I heard a seminar by Matt Scott describing the *Antennapedia* walk, working out the map of the region of *Drosophila*, and I said, "We've got to do this [in worms], but not just one bit—the whole thing!" Because at that point there was this absolute bottleneck with the worm people.

I remember discussing this with Bob Horvitz and Jonathan Hodgkin. And I remember being quite worried. It wasn't at all clear to me that the worm was going to make it! We lacked some of the perquisites of the fly, the polytene chromosomes in particular. The fly came with this built-in physical map, which the worm didn't. And one of the consequences was that our students were spending years, their whole PhD theses, isolating a gene. It was awful. John White complained about this. He said the seminars were deadly, mind-numbing to sit and listen to each student explain how they had failed to clone their gene.

So after hearing Matt speak, I came back and talked to people about how we're going to do this. And I said "Look, we're going to start." So I was given some space in the new bit at the LMB to do this.

Gitschier: Did you ever imagine that it would turn into what it eventually became?

Sulston: No, absolutely not. I just saw it as a specific problem and it was very much oriented on the map.

Gitschier: And the recovery of clones.

Sulston: Yes. Just sort the thing out and have the clones lined up in the freezer, so people could take them and have a reasonable correlation between the physical and genetic map. That was it—that was to be the project for the '80s.

That really was a big change in my life, and that led directly beyond, not only to the worm sequence, but then to the human. As night follows day, to this place being built. And everything that happened in the '90s around here.

Gitschier: Was Alan Coulson just leaving Sanger's lab at that time?

Sulston: That was the other very good thing that happened to me, and probably without that, it would not have worked. Alan chose to join me when Fred Sanger retired. I had already been working, rather ineffectively, on it. Alan is superb at making things work and parallelizing. So we had a hell of a fine collaboration and things started to move immediately when he came.

Gitschier: And it sounds as though you were engaged with this new project as much as you were with the old one.

Sulston: Exactly. It was an important problem. And we needed it [the map].

What I did for much of the '80s was make the libraries, for some reason, because Alan was getting on with the biochemistry. But the thing I mainly did was to learn computer programming. I took that up and wrote all the software. I had no intention of doing that, but there was no one else to do it, and we needed it urgently. So one weekend, I said, "I'm just going to start writing Fortran." Roger Staden started me off and it snowballed into this huge program, very unprofessionally written, but it worked.

When we had done two or three years of this and more or less exhausted the approach we had, which was looking at cosmids, we became aware there were gaps. Bob Waterston came and we all worked on this. From the mid-'80s this was a formal and equal collaboration. And it was resolved by Bob's going back and using the YAC [yeast artificial chromosome] technology that had just been developed in Maynard's [Olson] lab. And suddenly we had the whole map.

Gitschier: OK, let's fastforward to more recent history at the Sanger Institute. Why did you decide to step down as the director in 2000?

Sulston: Well, the decision was made in 1998, and the people I'd worked with weren't terribly happy about it—a bit rocking the boat.

We were just a bunch of amateurs, the seven of us on the board of management, and the staff did get very unhappy after about three years when the numbers went up to about 100, and the Institute wasn't being running along stable, robust, and predictable lines. So we had some management training that was quite interesting and extremely effective.

We all underwent a bit of psychoanalysis of our various characteristics. When it came to the results from me, the facilitator said "Well, you're the sort of person who gets there in the end by a very muddled route and you end up emerging backwards through the hedge covered in bits of glass." I was rather flattered; I thought he got that about right!

And that's back to your question about doing things differently. If you worry too much about how you're going to do something, you don't do anything.

Gitschier: So, back to leaving the directorship...

Sulston: It was like leaving the lineage work. I felt I had done what I could, and I felt quite strongly that it was going to plateau, and I wasn't a proper director, in the sense of really *enjoying* running things. I didn't have yet another big project that would move things on. It would become more of an administrative thing. What Allan [Bradley, current WTSI director] has done is to hire a lot of new faculty doing a lot of different things and the place has become much more like a university department with a lot of the high-throughput stuff still here.

I really can't run a department in this way. It was a single project thing for me.

Gitschier: Ah, you are focused.

Sulston: But I had enough of a sense of preservation of the Institute that I was always enthusiastic about having multiple things going on.

Gitschier: When you stepped down, did you want to relinquish the administrative responsibility but continue with the genome project?

Sulston: I didn't have a particular thing in mind. But at that point, there was still too much to do. In 1998 we had published the worm in incomplete form and still had a lot to do on closure. The human genome was just reaching its crescendo. And I told people I was not stepping down from that. I worked as hard as ever on the worm and on the human right through 2001, and then things began to fade a bit. And by that time I'd got the notion of writing a book with Georgina [Ferry].

So somehow, I never really considered what I would do next because the time was full.

Gitschier: I've also heard that when you win a Nobel Prize your time is no longer your own.

Sulston: It's very demanding. You have to sort it out. It's impossible.

Gitschier: Do you still have a lab here?

Sulston: No. A couple of years ago I became Emeritus with the Wellcome Trust. I pop in at least once a week, often on Sunday morning when it's quiet. Sometimes with a specific intent or a meeting, or I'll meet people in passing. I use the library quite often. I still like using books.

Gitschier: Looking back, was there one period that sticks out as being the most joyful or the most productive or the most intellectually exciting?

Sulston: I wouldn't discriminate. Something I want to emphasize is how distinct they are, really. My postdoctoral work on prebiotic chemistry with Leslie Orgel was hugely exciting, and before that I enjoyed being a student with Colin Reese just making some of the early oligonucleotides. But there is this sense that I, probably more than most, have jumped from one thing to another.

Gitschier: Yes, but with a rather slow periodicity. You couldn't be faulted for leaving a project before it was completed!

Sulston: Yes, a classic seven years. Well, I suppose the genome is rolled itself together into two lots of seven years.

Gitschier: When I read *The Common Thread*, I was struck by the image of you and the public human genome project as a ship under a state of attack. That you were thrust into a position of having to suddenly defend yourself and your approach against the assault of entrepreneurial javelins, and to work even harder. What a huge waste of time and resources the battle caused!

Sulston: It was hugely complex. And that was exactly why things changed again. And any immediate thought of restarting a group went out of the window! It was a waterfall of things happening! And then, all I wanted to do was to write an account of it afterwards and talk about it because I thought it was important.

But a lot of people say, "Well, this is the way the world is. You should know that!" And that I find very depressing. There is a sense in which we are struggling to come to terms with a very rapacious free market, in all sorts of areas. And this little particular area is what I was trying to express in the book.

There is a lot of rewriting of history now, but it was just like that—the assault with javelins.

What I've never been willing to agree to is that somehow or other it helped! The standard story, because everybody wants peace, is to say, "Well, yes this competition did accelerate things."

I think that is absolutely wrong. All it did was to speed up getting this *fake* release of the draft sequence, which was 90% complete. It was a political deal. It was an election year [2000]. The White House had really become unhappy about what was going on. It was a silly deal, but it meant peace.

Gitschier: Have you met Craig Venter?

Sulston: The last time I saw him was at the Gairdner. One of the things we've had to do since this is to all go and get awards together. What else can you do? You can't be in a state of perpetual warfare. And what you said about it being a huge waste of time is absolutely right.

Gitschier: Do you think Americans are most aggressive about privatization? Do you think such a thing could have happened in England?

Sulston: Well, we have rapacious entrepreneurs here, too, but America is bigger and has the world's richest corporations.

Of course, there was a battle within the US, which was almost divided down the middle—the neoconservative and Democrat wings. Yes, because the resources are there, I think these battles are more likely to be fought out in America.

But of course, increasingly so, they are also fought on the global campus, and that's the other reason why I think this is so important. We are wasting time, I think, to get globalization right, and people ought to be paying more attention to stories like this because it shows ways that things can be run stably in the future and ways in which they can't.

If you have unbridled competition as the basis for international relations, then we are going to die. Because we don't have the basis for sorting out the environment or people's lives or anything in a more ethically advantageous way. If everything is done through the World Trade Organization, and everything is done for who can get the most revenue, we're sunk.

Gitschier: Were you thinking, when you made plans to retire, that you were just going to end it [your career]?

Sulston: No, I knew I wasn't just going to end it, because I had too much to do. What I hadn't planned on was what it might lead to. I've dealt with how things have evolved and haven't started new ones. That was not a plan.

Gitschier: Can you articulate your role now?

Sulston: I am occupied with various small attempts to lubricate the interface between the science of genetics, human genetics, and the public—all of us.

I'm the vice-chairman of the Human Genetics Commission, a part of the Department of Health in the UK. The aim is to have a body of experts, about twenty of us in various areas—industry, science, nursing, consumer affairs—all people with a stake in genetics in general. The thing I'm most concerned with is genetic equity. Simply nondiscrimination, because it's a real issue.

Another thing I'm doing informally is speaking in support of various NGO activities in Geneva. I'm interested in fair trade. Medecins sans Frontieres and Oxfam both run campaigns on this which I've supported. There is the Drugs for Neglected Diseases Initiative, which I support.

There is also an active campaign going on in WIPO [World Intellectual Property Organization] to bring a development agenda. The background to all this is immensely complex. In the WTO [World Trade Organization], the TRIPS agreement [Trade Related Aspects of Intellectual Property Rights] is coming to a head and is forcing all but the very poorest countries to sign up for a high level of enforcement of law as practiced by the G8 countries. What this means is that India will no longer be able to produce generic drugs freely. It will greatly increase the powers of the major corporations.

One must not be cavalier and say, "This is all bad and we must sweep it away," but unquestionably the current system we have for funding medical R and D leads to production of pharmaceuticals only for the rich markets.

Gitschier: What about the Gates Foundation?

Sulston: The Gates Foundation is the *only* major source of revenue for these initiatives right now. And all the public/private partnerships are dependent on Gates money at the moment.

It's still not enough and it's not sustainable enough, even though Warren Buffett has just contributed his money as well. So perhaps the world will be run by rich men's charities. I find that rather uncomfortable. Of course, the reason we are here in this building and sequencing the human genome is because of pharmaceutical stock from the Burroughs Wellcome Foundation.

So things are a mess. What we're up against is whether we can have real intergovernmental support for drug development that will not depend on the marketplace. *PLoS Biology* [http://www.plosbiology.org] published a new concept for an international treaty on biomedical research. Jamie Love, who runs his own NGO called CPTech in Washington, and Tim Hubbard, who is here, have put together a scheme for an international accord by which countries contribute to a global healthcare R and D fund. This can be done through the free market if people wish, but equally could be done in an open access sort of way, through institutions. It would be much more transparent. Above all it would allow us to devote funding not just to the revenue-earning things in the rich markets. But we could start to address the neglected diseases.

It would generate an enormous worldwide fund, far bigger than the Gates Foundation. This is the kind of proposal that really raises the ire of the pharmaceutical representatives.

It would be a hugely cohesive thing in the world if we were able to move towards a global health system. But I fear that the force is going the other way. The things I've heard the representatives of the pharmaceutical manufacturers say in Geneva—the blatant, vicious manipulation of figures to prove their case! And it sounds silly—but just the rudeness!

Gitschier: The arrogance?

Sulston: Yes, exactly. They say: "You're wasting your time. Nothing will ever change." How appalling!

Gitschier: Maybe these kinds of problems make the human genome project seem not so complex!

Sulston: When I give speeches to people, I do tend to end up as a prophet of doom, saying we really maybe won't live another century. And one of the ways we may *not* do it is to continue along this path.

Gitschier: On a much lighter note, I have a final question. I will never meet a knight again, so...

Sulston: There are quite a lot of them about, actually.

Gitschier: There are?

Sulston: This is quite interesting. For a long time the guys in LMB didn't accept knighthoods. Max Perutz, for example. And not long before he died, I asked him about this. He said he didn't because it would set him apart from the staff.

Now, it's important to look at background on this. Those guys, when they set up the LMB in their new building, set up a single canteen. More or less contemporaneously, I was in research in the Chemistry Lab, and there were three canteens—one for the faculty, one for the students, and one for the technicians. Absolute class distinction, and that is the way it was always done.

So, in that context, Max didn't want anybody to be put in a position of uncertainty about how to address him. He was just "Max."

Max was not a sort of hale fellow. He was very quiet and quite formal, so people might have been inclined to give him his title. He was a very properly spoken kind of guy. And it was wonderful to learn that underneath was this well of egalitarianism.

Gitschier: Fascinating.

Sulston: It is a real problem with knighthoods.

Gitschier: Sorry, I must laugh, sounds like a problem with putting on your armor or something!

Sulston: The reason I said that there are quite a lot of them is because there are! It is not the top honor. You also have MBE, OBE, CBE, then knight, then various higher ranks.

Gitschier: You just jumped in at knighthood?

Sulston: Yes. People put you up for these things. You just get a letter one day saying you are being offered a knighthood, are you going to accept it? There is just this curious convention that with the knighthood you can put "Sir" in front of your first name, not your second name.

The whole thing is absolutely ridiculous, really. But I was persuaded that it was quite a good thing for science to be recognized in this way in this country. Aaron Klug was the first one to accept it at the LMB. And he began to talk to us about how accepting knighthood was good for science. Otherwise science was the poor relation and business and arts were recognized. Of course, some people ask how could I accept this—republicans and the [Manchester] Guardian—and I say this has nothing to do with the Queen. Although you do meet the Queen.

Gitschier: You do?

Sulston: Yes, the sword on the shoulder. It's all pretty odd. So we either cold-shoulder it, or we say scientists are just as good as anybody else.

Sweating the Details

An Interview with Jamie Thomson

Because any cell taken from the earliest stage of a mammalian embryo has the potential to produce a full organism, embryonic stem cells, in theory, could generate replacements for liver, skin, or even parts of the brain. Yet, application of this technology for humans ushers in ethical concerns as it involves destruction of an early human embryo. The originator of human embryonic stem cells discusses his work as well as more recent technological advances that can be used to convert cells from any adult tissue back into their embryonic state.

Interviewed April 10, 2008

Published August 29, 2008

IF YOU HAD TO NAME THE MOST CONTROVERSIAL SCIENTIFIC ACHIEVEMENT of the past decade, you'd be hard pressed to top the development of human embryonic stem [ES] cells. Human ES cells followed on the heels of another major technological advance—Dolly, the cloned ewe. Together, these remarkable breakthroughs have stimulated great public interest and have ushered in a new era in the exploration of human biology. At the center of the ES maelstrom is a soft-spoken and intensely private scientist from the Genome Center at the University of Wisconsin. Jamie Thomson, who is also Director of Regenerative Biology at the new Morgridge Institute for Research and the founder of two companies, is purposeful, with an obvious knack for a difficult experiment, yet seems a bit uncomfortable in the limelight his work has generated.

Trained as a veterinarian and research scientist, Thomson had a dual passion for experimental embryology and species preservation. He was just emerging from his post-doctoral fellowship at the Primate Center in Oregon when he moved to Wisconsin. He was hired there with the specific goal of deriving ES cells from primates, a feat he accomplished in short order. Within a few years, working closely with the ethics and in vitro fertilization (IVF) communities, he succeeded in deriving human ES cells. More recently, pushing the boundaries of human developmental biology even further, his laboratory reported the creation of induced pluripotent stem (IPS) cells, a way for turning human differentiated cells back into ES cells.

I sought out Jamie in mid-April during a cold, umbrella-inverting rainstorm. With daughter in tow, I borrowed my sister's red MINI Cooper and headed off from her home in Milwaukee to Madison. After struggling with the concept of paying for parking by phone, we puddle-jumped our way to Thomson's building and literally stumbled into a sign celebrating the first synthesis of DNA, one of a series of placards that extol the long line of important discoveries made at this University. I located Thomson in his office down a quiet hallway, abandoned my daughter to the tea room and her assigned reading in *To Kill a Mockingbird*, and to the soothing sound of the University of Wisconsin coal train outside his window, we began the interview.

Gitschier: The first thing I want to talk to you about is how you got into this business.

Thomson: In college, I liked biology, and I wasn't quite sure what I wanted to do with it. I was good at mathematics. I was trying to find a way to put the two of them together in a biophysics major. But I was spending a summer at Woods Hole, doing population biology stuff, because I thought that was a good overlap with understanding mathematics.

Gitschier: What kind of organism were you working on?

Thomson: I was working on the salt marshes with *Melampus bidentatus*, which is a little snail. And I was up to my knees in mud the whole summer in the Great Sippewissett Marsh, and I decided that I liked mathematics, but I wasn't really into counting things that large. So I was looking around for a way to put these things together, and I liked developmental biology still.

One night, I went to a lecture by a guy from Stanford at the time. His name was Paul Ehrlich, an ecologist type, and he gave a talk on endangered species. He quickly went through this argument that zoos would have no positive impact [on preventing extinction] because of the size and the numbers [of endangered species] involved. If you filled up the zoos with that number, you'd hit only a very small number of species, and you're not going to do anything significant.

I was sitting in the audience going "Hmm, I thought you could freeze things like sperm and eggs." I was a biophysics major, what did I know? But I thought that if you went out and collected enough sperm and enough eggs, you could keep fairly large repositories and the breeding problem could be managed artificially through this stock of frozen stuff.

That same evening on the news there was a woman named Barbara Durant with Walter Cronkite at the San Diego Zoo, and she had this rat that had come from a frozen embryo.

So all of this happened in one day! And I looked at that and said, "I'd like to do that! I wonder what kind of degree would be useful for that?"

Gitschier: What year was this?

Thomson: This would have been 1980, between my junior and senior year.

I had an undergraduate advisor named Frederick Meins who, even though I had a very good grad school advisor, has been my mentor in life. When I was 18, he told me all these wonderful stories about these biologists, one of whom was Jared Diamond, who would be on the cover of the *New York Times* for discovering some new species in New Guinea. But he was also a biophysics professor! And I thought, "Hey, I could do something like that!"

So the plan was, in my naive undergraduate ambition, to do practical veterinary things that would allow me to manipulate embryos of endangered species, and at the same time, do basic developmental biology. It turned out that Penn had a combined VMD/PhD program. So, I went to graduate school there in a basic mouse developmental biology lab [with Davor Solter] at the Wistar Institute.

When I went to be a post-doc, I decided to switch to primates, to go back to this original idea. There are about 200 species of primates, depending on whether you are a lumper or a splitter, and at that time, about half were endangered or threatened. And it seemed like a focused population with biomedical relevance. And if I did basic developmental biology in primates, it would be directly applicable to humans in a way that mouse is not.

So I went off to the Oregon Primate Center, which at that time had the best embryo recovery and in vitro fertilization [IVF] program for nonhuman primates. I learned a lot about doing IVF and experimenting with primate embryos and, at the same time, I initiated some fieldwork in Sulawesi [in central Indonesia] with an endangered primate called *Macaca nigra*. They had gone from 300,000 in the '70s down to about 3,000 by the time I got there. I was going to attempt to dart the remaining male population, collect semen, and bring it back home—some preliminary stuff to get a grant. I think the grant came in second and they [the NIH] decided to fund only one. So I started looking around for a job, because things were ending in Oregon.

I moved here [Wisconsin] because John Hearn was the director of the Primate Center here, and he had worked out methods to have in vivo–recovered embryos. He hired me specifically to derive primate embryonic stem cells.

Gitschier: Were these embryos the product of in vitro fertilization?

Thomson: No, these were natural ones, flushed out in a nonsurgical procedure. We got very high-quality embryos that don't require culture. The advantage [of using natural embryos] was that back then, the culture medium [for IVF-generated embryos] was pretty bad, and you couldn't go from one cell to blastocyst and have a real healthy product.

So I came here specifically to derive primary embryonic stem cells.

Gitschier: But wait, is this a leap? You were interested in saving embryos that could then be used to regenerate lost species, potentially.

Thomson: No, the intent was to establish a robust experimental embryology in primates that could make up for some of the species-specific differences in mouse. Mouse is simply a better model, there is no way around that, but in some ways it does not reflect human development very well. And in Oregon, despite the fact that they had the best embryo recovery program from IVF in the world, they were completely starved for embryonic material.

In mouse, you can sit down and do 200 embryos in a day—it's not a big deal. The cost here for the natural flushed [primate] ones was $2,000 per embryo! So you couldn't do a thousand of them, you couldn't do two of them! It was almost impossible.

So, the rationale for me to derive the primary embryonic stem cells was to get a sustainable primate material that would recapitulate normal human events better than mouse ES cells in a way that the material is not limiting.

Gitschier: OK!

Thomson: It's all about experimental embryology. And Hearn had been Director of the research branch of London Zoo. He had a strong interest in promoting conservation of primates when he came here. And I still had this idea that I could do both [experimental embryology and conservation], until I derived embryonic stem cells, and then it kind of took over my life.

Since 1995, when we published the first rhesus embryonic stem cells, those other interests in endangered species have been pushed aside.

Gitschier: Would you like to get back to that sometime?

Thomson: Part of me wants to. I have a young family now. Mucking through the jungle is not going to happen any time soon. Although it is attractive and I would probably enjoy doing that more than what I'm doing now.

Gitschier: Let's talk about 1995 and your work on primate embryonic stem cells. It sounds as though it worked fairly quickly. What did you do that was different from the mouse?

Thomson: It was sweating the details. Superficially it was the same. The primate cells are not dependent on LIF [leukemia inhibitory factor], but feeder layers worked also for these cell lines. We were very careful with the culture conditions and got it to work. Later on, we and others discovered that cytokines that mediate self-renewal are distinct between the two cell types, but we didn't know it at the time.

Gitschier: Why had no one done it before? Mouse ES cells were derived in 1981. Fourteen years later...

Thomson: In the middle 1980s, people in Britain had already tried to do human ES cells and failed. And failed probably because the embryos weren't very good, because the culture medium was bad, and they were still thinking they were like mouse ES cells, which they are not. Even though the conditions we use involve fibroblasts, as in mouse ES cells, the finesse is different, the timing is different, the splitting is different. Some very good mouse ES cell people attempted to do this, and I don't know why they failed, but they did fail.

We had good access to very, very high-quality primate material. We were very quickly successful with that, and because we had that experience, as soon as we had high-quality human embryos, we got it. If you look at the numbering of our cells lines, it goes from H1 to H14. H1 is the very first human embryo we tried, because of our experience with primate. And the culture medium was just getting better at the time we were developing our cells.

Gitschier: What was different about the medium?

Thomson: David Gardner, working in Australia and now in Colorado, developed a new generation of media; it was optimized for human material. If you put mouse media on human embryos, it doesn't work very well. It's nothing to do with a special growth factor, it's just optimizing salts and glucose, stuff like that.

Gitschier: So it's a medium for the embryos themselves, not the stem cells.

Thomson: Right. Prior to that we couldn't get a quality blastocyst to try it [stem cell derivation] on. With the primate work, we didn't need the medium because we went straight to the derivation process, because the [in vivo–recovered] embryos were high quality.

Gitschier: So were you communicating with the people who do IVF and who actually culture the embryos?

Thomson: Oh yeah. Jeff Jones was the person who actually did that and Gardner had a graduate student who became a post-doc at UW, and that post-doc helped Jeff introduce that medium into the IVF clinic. So this clinic was among the first to use that new generation of media. Part of the reason we were successful is that Jeff is very good at his job.

Gitschier: So, 1995, you were still interested in primates for primates' sake, really, looking at early embryonic development. Why did you then make the leap to humans?

Thomson: At the time we derived the primate stem cells, I really wasn't planning to do it. I assumed that once we published the primate work, someone would do the human work very quickly. We have a very small IVF unit here, with very limited access to embryos, and I thought that every IVF lab in the world would be doing this quickly.

Gitschier: And presumably, it wasn't even a primary interest of yours.

Thomson: No, though it was obvious that it was really important. I just thought someone else would do it. But as the months wore on and nobody did it, I decided to try it here.

Gitschier: And did you do this with your own hands?

Thomson: Yeah, I did all those experiments.

At the end of 1994/95, I spoke to our ethics people here. In 1994, Clinton asked [Harold] Varmus [then director of the NIH] to start a commission to look at embryo experimentation and Alto Charo, who is a lawyer here, sat on that panel. So I asked her, "Hey, if I were to do this, what should I do?" I was really lucky that she was here. Also, the head of our IRB, Norm Fost, was very supportive.

Things have basically been "won" now, from a public perception. But back then, everybody was really scared. They were scared about public funding. They were scared about personal safety. While the University wasn't ecstatic about having me around, because it made everybody nervous, they were all really supportive, and in particular those two people helped me put the consent process through in a very reasonable way.

Gitschier: Did you personally have any ethical issues with the human embryonic stem cells?

Thomson: I thought about it a lot prior to doing it, and I decided that if these are embryos that the patients have already decided to throw out, that it is a better ethical choice to use them for something useful. And it's actually a fairly complex problem about how you think about that.

Gitschier: At the time was there already some kind of prohibition to funding human embryonic research?

Thomson: Yeah, it was actually worse then. The Dickey amendment [1996] said something to the effect that no federal money can be used to damage or jeopardize a human embryo.

Gitschier: So in addition to getting the IRB approval, you also had to get money.

Thomson: Right, and that's what made the University so nervous, because they would lose all the federal funding if I screwed up somehow.

Gitschier: No pressure! Is that when you started to be supported by Geron [Corporation]?

Thomson: I asked the University for about $20,000, and they said "No." And the next week Geron walked into my office and I said, "Well if the Federal government won't fund me and the University won't fund me, great!"

Mike West was the fellow at Geron at the time. He's got the vision thing down. He understood that this was important before other people did. I accepted funds from Geron until President Bush made it legal to use federal funds, and I have declined them ever since.

Gitschier: But you need to use existing cell lines.

Thomson: I've used the same five cell lines for 10 years now.

Gitschier: You made a comment in another interview: "I didn't know we'd be stuck with these cells."

Thomson: Yeah, I assumed people would derive new ones right away.

Gitschier: But the implication is that these early embryonic stem cells might not be as good as ones that might be derived later.

Thomson: Actually, to be stronger than that, I very specifically did not keep track of stuff, like lot numbers, because I didn't want to have to do the paperwork for FDA-like stuff. I figured, this is nice proof of principle. We'll just derive MORE! If people want to do a GMP [good manufacturing process], let them. The FDA actually wants tracking now.

Gitschier: Let's talk about how your life changed then in 1998 with the publication of human stem cells.

Thomson: Well, Dolly had been cloned the year before, so I was kind of prepared, because I could see what Ian Wilmut had to go through.

Gitschier: And what did he have to go through?

Thomson: Hell. It doesn't allow you much time for doing your work anymore.

So I was wrong about some things. For one thing, the media were pretty positive, but it wasn't clear prior to publication how it would go. On the whole, the initial science reporting especially was superb and the story was accurate.

The other thing I miscalculated was that I figured people would have a pretty short attention span. But we're 10 years in, and it is still a big story. I think it has to do with politics and who got elected to the White House more than anything, because I think had somebody else been elected, it [stem cell research] would have been "normal" science a long time ago.

Gitschier: And now, you've developed a new technology that may obviate some of these political and funding issues: induced pluripotent stem [IPS] cells. Set the stage for that for me.

Thomson: The stage was Dolly really—that changed the mindset of developmental biologists in a big way, including mine. About 5 years ago, I hired the post-doc [Junying Yu] who was the first author on our paper [published in 2007]. My conversation with her at the time was that we have to try this, even though it probably isn't going to work. And it's probably like a 20-year problem, because the thought back then was that it has just got to be really complicated. All those little factors, and how can you manipulate all of those? It didn't really seem sensible.

I thought by doing such a combinatorial screen, we might get *partial* reprogramming in some way.

Gitschier: Describe what you mean by combinatorial screen.

Thomson: I'll tell you what we did, and it was very similar to what Yamanaka did in the mouse. We were doing it at the same time, but he got ahead of us because mouse work is actually much faster than human work, although we actually had a partially defined system with a more complicated set of factors prior to publication of his mouse work.

Back in the '70s, it was found that if you fuse blood cells with embryonic carcinoma [EC] cells—ES cells hadn't been derived yet—that within that heterokaryon, the dominant phenotype could either be the blood cell or the EC cell, but it was often the EC cell. So that was early evidence for reprogramming.

We started to do similar experiments several years ago, in which we took ES cell–derived blood cells. We had a well-defined, cloned, expandable hematopoietic cell type that we used in cell fusions for a model for reprogramming, and we showed that the dominant phenotype was the ES cell.

We did gene expression analysis of both those cell types and started to clone genes that were specifically enriched in ES cells. So Junying cloned between 100 and 200 genes, and she started taking pools of them to test for reprogramming ability and we used a knock-in human ES cell line that turns green and gets drug resistant when it reprograms to an ES cell state. Last summer, Junying kept paring it down until there were four factors, and we repeated it in different cell types.

It was kind of a dumb thing to do—it worked and that is nice. If you look at the factors we found, *OCT4*, *SOX2*, and *NANOG*—they're everybody's favorite genes already—these are key pluripotency genes. But we had this mindset, which was so strong, that it HAD to be complicated, we just never tested them! It would have been a lot easier to just test them 5 years ago and gotten it done in a month or two!

Gitschier: These IPS cells won't be restricted in terms of federal funding?

Thomson: No. That changes everything.

Gitschier: Do you think people will work both on IPS and ES cells?

Thomson: For our lab, we've been growing the same ES cells for 10 years—we're not going to stop that anytime soon. On the other hand, as new people enter the field, I would guess that most of them will be deriving IPS cell lines and not ES cell lines, and that over time, ES cells will take a smaller percentage of people's attention. It comes down to whether they are equivalent or not, which we don't know for sure yet. But my sense is that they will be. If you can't tell them apart, why inherit all the baggage of the embryonic stem cells when you don't have to?

Gitschier: So, what are you going to work on now?

Thomson: I'm interested in whether this is the first example of many. There are other cell fusions between other cell types, and it suggests that other similar lateral transitions could be artificially induced. Nobody has done systematic screens for those lateral transitions in a similar way, probably because everybody thought that was too complicated. There is probably no clinical utility for making a heart cell out of nerve cell, but that is the kind of thing we're going to see. Can we understand enough about the biology that we can predict which factors are needed without doing these combinatorial screens? I don't know precisely how you get this into regenerative medicine, but when you actually think about the nuts and bolts of how you'd introduce something into a patient, that is challenging.

The long-term goal of regenerative medicine is to cause tissue to regenerate, not to do cell transplants. And for tissues to regenerate that don't normally do so, you're talking about changes in cellular states that are physiologically disallowed. So if you understand what creates those barriers and how you can overcome them, it could well lead to real robust regenerative medicine. And that one is *definitely* a 21st century problem that won't go away as fast as the last ones.

Gitschier: Are you having fun?

Thomson: Yes and no. It's very satisfying—that what I do I think is important. But day to day, most jobs are just stressful, you know? Even if you are good at it, and things are going well.

Gitschier: Do you think that part of the stress, though, is the nature of the problem itself and the competition?

Thomson: Yeah, the competition has gotten—I won't say out of hand—that is the nature of the business—if something is perceived as important, there will be competition, but it clearly makes it a lot less fun, and I don't see a way around that. Especially the reprogramming stuff over the last year or so—everybody is doing that now! That was one of nice things about doing the primate ES cells—nobody cared!

By nature, I am a loner. At the last ISSCR [International Society for Stem Cell Research] meeting, I don't know how many people show up now, but based on my personality, too many! I do better one on one or in small groups.

Gitschier: But this is a path you're going to be on for a while.

Thomson: Yes, and I'm looking for little niches that I can find fun again for which there is not a head-to-head competition, 'cause at the end of the day, if you publish a week or two before somebody else, it's kind of futile, isn't it?

Gitschier: Final thoughts?

Thomson: The point that I want to make is that a lot of the enthusiasm and emotion that has driven human embryonic stem cells is the idea that we'll use this for transplantation and cure diseases like Parkinson, and while I think in limited cases that might be true, I think broadly it will prove extraordinarily challenging. And if you look at the field say 10 to 20 years from now, there'll be some stellar successes in the transplantation realm, but there'll be a lot of failures, and I think people are ill-prepared for the risks of a new technology like this.

If you look at the early days of bone marrow transplants, most people died! I think a single death in this area is going to create an uproar, given the politics involved, and there will be such deaths, because the diseases people are contemplating [treating] are very serious diseases.

So if you look at the pie of things that will be important 20 years from now, my belief is that the broader applications are in the understanding of the human body, and that, similar to recombinant DNA, it will be pervasive and everybody is going to use it and they won't call themselves "stem cell biologists" anymore. It'll just be something to get access to the human body. And I think that *will* profoundly change human medicine in ways that I can't even predict.

The Making of a President

An Interview with Shirley Tilghman

Mammals are diploid organisms, but the genetic information contributed by the mother and father is not always equivalent. Occasionally, only the paternally or the maternally inherited allele is active, due to an epigenetic phenomenon known as "imprinting." While primarily discussing this occurrence with the president of Princeton and the particular obstacles faced by women in science, this interview also considers an intriguing example of an imprinted gene known as H19.

Interviewed on February 22, 2006
Published June 30, 2006

TO SEE ONE OF OUR OWN AT THE HELM of a major university is unquestionably inspirational. And for those of us with two X chromosomes, perhaps even more so. I tried to imagine myself in the role of laboratory Cinderella: plucked from the work-a-day world of robocyclers and biohazard waste to a realm of literati, artists, and trustees, gliding into, if not glass slippers, pantyhose and pumps and a far spiffier wardrobe, my voice transformed from the constrained format of a journal article to one that can carry sway in every aspect of our culture.

This transition must have been second nature to Shirley Tilghman, President of Princeton University. She is the model of grace, intellect, perseverance, womanhood, and tact. Just 24 hours before my interview with her was scheduled, Lawrence Summers (President of Harvard) had resigned, and I half anticipated that I might be bumped for a sound bite. Indeed, her phone was ringing off the hook, but our interview held fast.

I had driven to New Jersey from my father's house in Pennsylvania the night before. As I turned north on Highway 206, I took in the Lawrenceville playing fields on the right and the stately old homes on the left, recalling a day 36 years ago when I followed the Orange Key Tour, the last time I had visited Princeton. I recognized Nassau Hall, the pair of bronze tigers protecting a set of massive, deep-blue doors embedded in ivy-covered stone walls. I approached the old building, walked across the marble foyer to an unmarked door, and fell into a cozy warren, made even more inviting by soft green colors and a bowl of chocolates.

Shirley met me and led me down a little set of steps to her office. Journal covers prominently displayed near her desk remind her of her former life, as does a bronzed Rainin pipetman with a plaque that reads "I'm a genius!", given to her by her lab when she became president. I told her the topics I wanted to cover in our interview.

Gitschier: ...and third, I want to talk about Princeton!

Tilghman: Yeah!

Gitschier: So why don't we start with that? I have read a couple of interviews with you, and I was so impressed by your comments about how you just love the institution. You've been here for almost 20 years and president for almost five. Tell me, why do you love it here?

Tilghman: There are really two answers. First, it's a place that is always striving for excellence, and excellence with integrity. I've loved that since the day I arrived.

The second is that it is a place that is run by the faculty. Here I am, as testimony to that! In fact, if you look at the history of the university, there has been only one president in recent history that did not come from the faculty. It is a place that respects scholarship, respects intellectuals, and believes that a university is best run by people who grew up caring about the life of the mind.

Gitschier: That's wonderful!

Tilghman: Yes, and the third thing I love about Princeton is that the students are just spectacular. They challenge you and make you better. Interacting with them in my old life in the lab and in my new life in Nassau Hall, every day is just fun.

Gitschier: How many students are there?

Tilghman: There are now about 4700 undergraduates and we're going up to 5,100. We're in a process of expansion, as a matter of fact.

Gitschier: Why do you feel the need to increase the undergraduate population?

Tilghman: There were a number of reasons, but probably the most important was that during the last 30 years, we had kept the population of the undergraduate student body relatively constant, while we were growing the faculty at a rate of about 1% per year. And we had reached a point where the student-faculty ratio was under six, which is extraordinary for a university, particularly a research university.

Second, we have the largest endowment per student in the country. I think that is a joy, to a president, but it also creates a responsibility. I believe that the trustees, who ultimately made the decision to expand, believed that if we have the capacity in both faculty and resources to educate more students, we should.

Gitschier: And the graduate student body?

Tilghman: We have about 2100 graduate students. The graduate school student population has always been roughly a third the size of the undergraduate student body. We can do that because, of course, we have very few professional schools.

Gitschier: Do you think that there would be some virtue to having a law school or a medical school affiliated with the university, or is that just not brought up?

Tilghman: It comes up with a periodicity of about 25 years.

Gitschier: Pretty slow!

Tilghman: But it is studied periodically. The trustees cogitate about it, and, at the end, they decide that one of our distinctive characters is that we are, as an institution, so focused on two things: (1) the best quality undergraduate education that we can deliver, and (2) really stellar PhD production. The fact that we're so focused is one of our strengths.

Gitschier: I read one account of how you had been on the search committee [for the president] and then, while you went off to teach a class, the other members decided to offer you the job!

Tilghman: It was more, "Shouldn't we think about the possibility that she might be a candidate?" They were still six weeks away from making their final decision, but that is roughly how it happened.

Gitschier: I'm thinking of how weird this is, to go from being a professor with almost no administrative experience....

Tilghman: What were they thinking?!!

Gitschier: What do you think they saw in you? Did you have a vision that you articulated to them?

Tilghman: The search committee met a lot for the first four months. We spent a lot of time talking about the presidency and about Princeton. We traveled together, in pairs, all over the country interviewing other presidents, leaders in the scientific community, and those who would have perspectives on what a university is going to face in science.

So this group really knew each other well by the time they asked me to step down [from the committee]; we had all been very frank with one another. This process was highly confidential, so people felt free to express their views. I suspect that by the time they asked me to step off the committee, they had a pretty good idea of how I thought about Princeton, what the challenges were likely to be, what I thought we should be looking for in a candidate. I had supported some of the candidates who were clearly rising to the short list, so I think they also had a sense of my taste.

Gitschier: At the time, what did you think were the challenges, opportunities, and future directions for Princeton?

Tilghman: Certainly, at the time I felt strongly that it would be helpful to have someone who could grapple with some of the challenges that research universities face—to sustain what I think is our critical role as innovation engines for the country. And as we look at what has happened in Washington over the last four years, I think this concern has been borne out—despite what we heard in the State of the Union address—"Show me the money!" I was worried that as science became more expensive and more complicated, how could we continue to provide support in universities? After all, there is nothing sacrosanct about the idea that fundamental research must occur in universities. It could have easily been decided in the late '40s that federally funded research should occur in research institutes or in national labs. I think, though, the connection between science and teaching has been so productive for this country.

The other thing that I was concerned about on the search committee was how a place like Princeton, which has both the joy and the burden of great tradition, balances respect for that tradition with the need to move forward.

Gitschier: So, how do you like your job?

Tilghman: I love my job!

Gitschier: What do you love about it?

Tilghman: Well, it goes back to your very first question. I love this place. You can effect change in real time. You can see the effect of what you're doing, from recruiting a senior faculty member down to simple things like hearing about some ridiculous rule and changing it to make everybody's life easier.

Gitschier: Power has its merits.

Tilghman: You have to be careful, of course, because you don't want to become a rogue elephant.

Another reason that this is such a great job, which I didn't fully anticipate, is the opportunities it has given me to broaden my intellectual horizons. For example, I was in New York City yesterday talking to the Commissioner of Cultural Affairs about how our new arts initiative can engage New York City more effectively. This is an area of the university world that I knew very little about until I became president.

Gitschier: Do you ever feel overwhelmed by all the different tasks you need to attend to, or is it manageable?

Tilghman: It's manageable, because Princeton is so focused on trying to be very good at a small number of things.

Gitschier: You are still in the molecular biology department?

Tilghman: As far as I know! They haven't disowned me yet.

Gitschier: You don't teach any more, do you?

Tilghman: I do teach! I teach part of the introductory molecular biology class to freshman and sophomore students. And I advise senior theses and junior independent work.

Gitschier: So, you still have a laboratory!

Tilghman: I just sent off my last paper to *Genes and Development*, which I hope will publish it. My last post-doc is about to leave, and then I'll close the lab down.

As you can imagine, it is not possible to run a lab and run this university at the same time. That was one of the clear implications of agreeing to do this [job].

Gitschier: When I've told other women I am about to interview you, they say something to the effect of "Oh, she's my role model—have you heard 'The Speech'?" I've never had the benefit of hearing you speak about women in science. Perhaps you could condense it into a few minutes of tape! Or perhaps I can ask a few questions and maybe the speech will come out.

Tilghman: Let's try that!

Gitschier: OK, let's start then with your early career. You were an assistant professor at Temple [University] and then at Fox Chase [Cancer Center] in Philadelphia. Were you then looking at Princeton because of the proximity?

Tilghman: No, I began to think about leaving Fox Chase when David Baltimore tried to recruit me to the Whitehead [Institute], and I looked at it very seriously. I saw a great deal that I found enormously attractive, including the chance to be with Rudi Jaenisch, who is in my field and whom I adore. It was after it became known that I was looking that Arnie Levine called and asked if I would also look at Princeton.

At the end of the day, it was really a family decision to come to Princeton. I was single by this point, with two little kids. As I tried to figure out how I could afford to live in Boston, send the children to private school, commute to the Massachusetts Institute of Technology (MIT), it just didn't compute. Princeton made everything possible—a little town with really good public schools, everything was within three minutes: three minutes to home, three minutes to the pediatrician, three minutes to the primary school, the nursery school. And it was a great university.

Gitschier: How did you manage? Your children were little.

Tilghman: They were age four and six years.

Gitschier: Did you have someone live in your home?

Tilghman: No, because it wasn't big enough. Over the years, I had a whole series of solutions to the problem you are identifying. If I had to do it again, I would definitely hire a live-in person. When I look back on the past, I think it was a terrible mistake not to do it.

Gitschier: I'm sympathetic to your history because I am also a single parent. We make these decisions so that we can function.

Tilghman: It's really true. If I had been making that decision purely on the quality and critical mass of science, I would have gone to MIT. You do what you have to do. And I've never regretted it for one minute. Princeton was exactly the right place for me at that time. And this has proven to be the right place for me in the long run, too!

Gitschier: Have we started to touch on "The Speech"? Are we getting close??

Tilghman: One version of "The Speech" is on the Princeton President's Web site (http://www.princeton.edu/president/speeches). The one that comes closest to what people are talking about is the one I gave at Columbia [University] last year about this time [March 24, 2005].

Gitschier: What was the topic?

Tilghman: It was about the future prospects for women in science—why it's important that women are represented in science, why it hasn't happened until now, and what we need to do to change things.

Gitschier: One thing that I've found incredibly helpful has been to have Howard Hughes Medical Institute (HHMI) funding.

Tilghman: Yes, HHMI was tremendously good to me. I wouldn't be studying genomic imprinting if it hadn't been for its support.

Gitschier: And why is that?

Tilghman: I was studying this *H19* gene, and everyone thought I was completely crazy. If I had had to write an National Institutes of Health (NIH) grant to defend it in 1985, I probably would have had a really hard time. The HHMI money, as you well know, allows you to work on whatever you want to work on.

Gitschier: Since you brought it up, how did you get interested in *H19*?

Tilghman: Because it was a mystery!

Gitschier: It was so different from what you had been doing previously, working on alpha-fetoprotein (AFP).

Tilghman: Yes, but there is a connection. My first real foray into pure mouse genetics was to study two trans-acting loci that regulated the levels of postnatal levels of AFP during mouse development. One recessive mutation caused a significantly elevated level of AFP mRNA after birth compared to wild-type strains. The question we asked was whether this trans-acting locus, called *raf*, regulates genes other than AFP. So we did a 1980s version of microarray analysis. We looked for genes whose hepatic expression declined after birth in a strain-specific manner, in a manner parallel to AFP. *H19* came out of that screen.

Remember those little Millipore filters that had grids on them and you'd grow cDNA-containing bacterial colonies on them and then hybridize them? Grunstein-Hogness! So that's what we did. *H19* was the H row, 19th spot. We had made a liver cDNA library and we were screening through them.

H19 was incredibly abundant and we showed that it was developmentally regulated, just like AFP. So this wonderful student, Vassilis Pachnis, who's gone on to be very successful at Mill Hill [MRC National Institute of Medical Research] in London, sequenced it, before machines. He kept hitting stop codons, and I kept saying, "Vassilus, you're making sequencing mistakes. It must have an open reading frame. There is no precedent for this for a spliced polyadenylated RNA." So this poor guy went back and continued to sequence and finally persuaded me that this was a highly abundant, spliced, capped RNA with no reading frame. And that was the mystery!

I kept plugging away on it, thinking there must be something here. Given its abundance and its tight regulation, this can't be a garbage RNA.

Gitschier: Well, I can see why you thought the NIH wouldn't fund it, because we are so entrenched in dogma.

Tilghman: Right. That is might be imprinted was just a lucky guess in the early '90s. It was one of these funny moments in science where you just have a leap of faith. You connect things and have a "eureka" moment. When I guessed it might be imprinted, I asked Marisa Bartolomei to check it out. It was the third gene shown to be imprinted, after insulin-like growth factor 2 and its binding protein, Igf2r.

Gitschier: Those genes are adjacent.

Tilghman: That's what got me going. At the end of the day we knocked out *H19* and there was no phenotype. If I have one regret about my career, it is that I had to stop before I figured out what *H19* does!

Unfortunately this little filter array name [*H19*] has stuck on it.

Gitschier: At least it's not an acronym that stands for something you can never remember, like so many other mammalian genes.

Tilghman: Exactly. I always think the fly guys had it right. Give genes very distinctive names.

Gitschier: I agree, but as a human geneticist, when you have to tell parents, "We're sorry but your child has a mutation in sonic hedgehog," there's something that seems weird about that to me. It sounds so trivial for something that has such serious import. But the spirit of the *Drosophila* people is fabulous!

But let's go back to your post-doc, working with Phil Leder.

Tilghman: It was a wonderful time. The NIH was the biggest sandbox I had ever played in. At that time, it was so exciting. There was literally no area of life sciences that you could become interested in and not find somebody at the NIH who was knowledgeable. So it was exciting and intellectually engaging.

And Phil was heaven to work for. He was very supportive and yet gave me independence. Cloning the globin gene, as you can imagine, was so exciting and then discovering that its structure was so different [having intervening sequences] from what everyone had anticipated was thrilling. It was one of those "hairs on the back of your neck" moments when you know you've discovered something important.

Phil has been a model mentor to me ever since. There is no turn in my career at which I haven't consulted him—including whether to take this job—and when he hasn't offered assistance, advice, moral support.

Gitschier: You are lucky to have that. Very few of us do. I think we need something to change so that women don't say "I can't do this" even before they give it a shot. Have you been able to affect the work culture at all?

Tilghman: I've increasingly started to worry that we spend too much time talking about how difficult science is and not enough talking about the fact that despite the difficulty—and I would never say it is not difficult because it is difficult—women can succeed and thrive. They can balance successful science careers and children. Think of the number of women that we know in that category.

Too often, I think, our graduate students and post-docs only hear the sob stories. They hear that it's so hard and impossible. I think if you hear that long enough, you inevitably become discouraged and conclude that you can't be so special that you'll be able to overcome all these barriers.

Gitschier: Who do you think they are hearing this from?

Tilghman: I think they are hearing this from us!

Gitschier: Aha—this is where "The Speech" comes in!

Tilghman: Yes. In the Columbia speech, I began with a whole series of pictures of extremely successful women in science who have had children and are now luminaries. And I ended with young women in science, just post post-doc, heading off to what I think are going to be the same kind of careers. I tried to make the point that it's not that it isn't hard. It is hard! It's hard for women to be doctors and lawyers. The more I learn about it, the more I think there is nothing unique about science that makes it more difficult for women to have a career than any other field.

It is possible. What we should be talking about more are the things you and I were just talking about. What do you need to have in your life to make it possible to have a career and a fulfilling life outside the lab? How do you organize childcare? How do you take maternity leave and keep the lab moving along?

Here at Princeton, we just created a backup [child] care benefit for all employees and students. You can call this company at a moment's notice and say, "It's a snow day tomorrow, I've got to have someone at my house at 8 o'clock in the morning."

Gitschier: Oh, that's great!

Tilghman: Isn't that great? And it costs $4 an hour. We are subsidizing it. But imagine the pressure it takes off people!

Gitschier: You must have very little time to relax now and your evenings must be booked!

Tilghman: Yes, that's one thing that has been very different. It's the nights, the dinners, the receptions. On the other hand, I often end up going to a student play or a dance recital or a basketball, lacrosse, or soccer game, and that is fun! These students come here and they entertain me! This is true! I don't find that it is stressful at all.

Gitschier: Of course, your children are grown up now. And that is hugely different.

Tilghman: Hugely different! I wouldn't have taken this job ten years ago. Couldn't have done it. Part of it is managing your limits. I had a little sign above my phone in the Lewis Thomas Lab that said "Just say no!" It's the only thing I've agreed with Nancy Reagan about.

Gitschier: Well, I'm awfully glad you didn't say "No" to me!

Turning the Tables

An Interview with Nicholas Wade

While scientists publish their findings in scientific journals for the edification of other scientists, it often takes a third party with good radar and a great way with words to sense the significance of a discovery and convey it to the broader public. This interview investigates the process of science journalism within the realm of genetics, as practiced by the New York Times.

Interviewed June 14, 2005
Published September 30, 2005

FOR MANY OF US, KEEPING UP WITH THE LITERATURE poses a never-ending challenge. We rely on word of mouth, "news and views," journal clubs, or e-mail alerts to stay abreast of our field. But the media are also potent arbiters of scientific advances. And in the realm of genetics, I can think of no better source than Nicholas Wade of the *New York Times*. What I like about reading Wade is that he gets right to the point—describing the discovery at hand in the first paragraph, yet roping me in for the full story. He's clear, crisp, and rarely misses the mark.

When visiting my father in rural Pennsylvania recently, I had the opportunity of interviewing Wade in situ at the venerable *Times* office building in Manhattan. The bus deposited me at the Port Authority Bus Terminal at the corner of 8th Avenue and 41st Street in blistering heat. I quickly walked two blocks north and turned the corner at 43rd Street to encounter the massive *Times* structure looming above, its signature globe lights forming a beacon to a set of revolving doors. I entered beneath the motto "All the news that's fit to print," registered at the front desk, and took the elevator up to the fourth floor, where I was met by Wade.

Wade gave me a quick accounting of the enormous newsroom, which at that level, houses science and arts reporters and editors. On passing through a maze of cubicles, he commented that it looked just like any office building. "A lot messier," I rejoined, as untidy stacks of papers were strewn everywhere. Inside a small conference room, I fiddled with my tape recorder, cognizant of a pro watching this neophyte. I looked across to a British man with a soft voice, lively blue eyes, and a puckish grin. He wasted no time turning the tables on me by asking, "Will the questions get too pointed?"

"No, they won't," I replied. "I'm very sweet."

"Your first mistake," said he. And thus we began.

Gitschier: How do you describe your beat?

Wade: We're a small department, and the boundaries of the beats are flexible. By and large, I cover whatever I'm interested in. I manage to keep out of others' way because I keep close to the frontiers of biological research, particularly genetics and molecular biology, and no one else has quite the same interests. Many of my colleagues write about medicine, for example, but I write only about things that are perfectly useless, given that it takes some time to translate basic research into anything practical.

It's one of the challenges we have as a science section—to get people interested in things that are of purely intellectual consequence.

Gitschier: How do you envision your readership?

Wade: As a general policy, the newspaper is addressed to the intelligent and informed reader, but it's always with the idea of bringing news. We're not in the business of education. People can find general information from a dictionary or the Internet.

For the science section, although we should have the same readership as the main newspaper, I assume that readers have a certain amount of scientific knowledge or interest, and, of course, many readers of the section are scientists. So we can sometimes put in more technical detail than we would in the main newspaper.

Gitschier: The *Times* doesn't do a readership poll to ask how many people are actually reading Nicholas Wade's articles?

Wade: Our business office does do those polls, but they are kept secret from us. That's with the idea that the content of the newspaper should not be driven by polls or market results but rather by what the editors think is important.

Gitschier: How long have you worked for the *New York Times*?

Wade: Longer than I like to think! I came in 1981.

I was on the news section of *Science*, and before that with *Nature*. At both journals, I was mostly concerned with political stories that affected science. I found I had a great deal to learn when I came back to science writing.

Gitschier: And before that, were you a scientist yourself or a journalist in another area?

Wade: I guess I always wanted to be a writer, but I was interested in science, so I read science at university [Cambridge], though without any intention of becoming a scientist. I didn't plan to be a science writer either, but the two things came together. I got a job with *Nature* in London when I was quite young, and was sent off to the Washington office that *Nature* had just set up. After a few years, *Science* asked me to join them. I thought it would be fun to work with an American company. I worked for them for about ten years.

The *Times* asked me to join them as an editorial writer to cover science, technology, and medicine, and I did that for about ten years. Editorial writing is great fun, but it's rather a limited art form.

Gitschier: Can you describe editorial writing?

Wade: Editorials are unsigned pieces [on the editorial page] because they are intended to be the voice of the paper. That anonymity may give the writer's words extra authority, but the disadvantage is that you lose your byline, unfortunately. So as a writer, you essentially disappear from public view.

The only exception is when you advocate a position that the editors do not think should be the position of the paper. The piece is then called an editorial notebook and appears with your byline to make clear it's just your opinion, not theirs.

After writing editorials for ten years, I became science editor. That, too, is a job that deprives you of a byline because there is almost no time to write. The science editors handle both the science stories in the daily paper, and those in the weekly science section. As an editor, you get to see how the paper works, which is of great interest, but you cease reporting, and spend a lot of time improving other people's stories and attending meetings.

One of the great things about being a writer, particularly an editorial writer, is that you get to know a lot. The whole world comes through New York, and many people want to talk to the *Times*. Often the reporters and editors on the main paper are too busy to see them, so they end up talking to the editorial board, where the pressure of work is much less. So even if you don't cover foreign policy or defense, you can get to meet the leaders in these fields by sitting in on your colleagues' meetings.

But when I became an editor, I found I was one step back from the front line of the news. Being unable to research or write anything, my intellectual capital dwindled fast, until I began to feel I had gone from knowing almost everything about the world as an editorial writer to knowing almost nothing [as an editor]. On a newspaper, the most interesting job is reporting. I went back to writing as a reporter five or six years ago.

Gitschier: I'm interested in the process that you undergo in developing a story. First, how do you discover what's out there?

Wade: Mostly, we find out through the main journals that we watch. Most of them have now become sophisticated in preparing what they call "tip sheets," or weekly lists of their most newsworthy articles, which are seen as a marketing tool for journals. They'll say to potential authors, "Send us your paper, and we'll get you mentioned in the press." Tip sheets are useful but insidious because it's easy to rely on them too much and not read the stories in the rest of the journal. So it's an imperfect system.

Gitschier: What journals give you these tip sheets?

Wade: A lot of journals do it now. I look carefully at the *Nature* journals, *Science*, *PNAS* [*Proceedings of the National Academy of Sciences*], *American Journal of Human Genetics*, and the *Cell* journals.

Gitschier: And *PLoS*, of course!

Wade: PLoS, of course!

Gitschier: Are you ever strong-armed by anyone to write about his or her work?

Wade: People sometimes do call up, but not as much as I'd like. Scientists are reticent about promoting their work to the press because they risk being criticized by their colleagues for doing so. But sometimes people call to say, "I've got something very exciting," and send me the paper in advance. It's always very useful to hear from people when they are enthusiastic about a result.

Gitschier: Do you attend scientific meetings?

Wade: Yes, I do, but not as many as I'd like. When you go to a meeting, you're usually obligated to write a story about it, and many scientific meetings are very hard to write about for the general reader because the findings are often incremental advances and difficult to summarize. On the other hand, meetings are very useful for talking to people in person, so I try to go to as many as I can.

Gitschier: After you read a paper, what's your next step?

Wade: I usually start by talking to the authors, and then call others in the field to see if they share the author's interpretation of the finding. Much of this can also be done with e-mail. I try to keep talking with people until I feel I understand a paper and its strengths and weaknesses, and then I'm ready to write it up.

Gitschier: Do you do most of your writing at home?

Wade: It's more restful to write at home. But if I have a story that will appear in the paper the next day, it's usually easier to be in the office.

Gitschier: How many articles do you write per week?

Wade: Usually about one or two, or more if there's lots of news. I've been on book leave for much of this year.

Gitschier: What is your new book about?

Wade: It's on what genetics is telling us about human evolution, human nature, and prehistory. I'm trying to integrate information from the many different fields that bear on the human past—paleoanthropology, archeology, historical linguistics, and evolutionary psychology—all of which are now being informed and amplified by genetics.

Gitschier: Back to the process. What's next?

Wade: You have to sell a story to your editor. It's a quite small department, so if a reporter says, "This is an important subject," it will probably go into the paper. But the question is at what length because space is at a premium. The editors who run the main paper, who tend to have a political/foreign affairs background, may not be as enthusiastic about science as we are. The science department's editors have to assess how much space they are likely to get for a story.

Gitschier: I want to learn more about which stories you choose to develop. When you look back on the stories that you've written about over the past five to six years, which ones leap out at you?

Wade: The first that comes to mind is the race to sequence the human genome. It was a good science story, but it was also of interest to general readers because of the rivalry between Celera and the university people.

Another story I found interesting was the genetics of human dispersal. We've had the picture of human origins as developed by the paleoanthropologists, and it's wonderful how well they've done with the material they've had available to them—just a handful of skulls. Then, the geneticists arrived on the scene and added a whole new dimension.

For example, there was a paper by Mark Stoneking about when man first started to wear clothes. He managed to figure out the date at which the human body louse, which lives in clothing, evolved from the human head louse, which lives just in hair. And the date of that divergence must give the time at which humans first started to wear reasonably close-fitting clothing. It's wonderful that genetics can provide that quite surprising insight.

Gitschier: My favorite recent story is *Homo floresiensis*.

Wade: Isn't that nice! The referees took a whole year, I think, to convince themselves that this was real. They started off by thinking this fossil must be a pathological *Homo sapiens* skull, but then realized it doesn't look like *sapiens*, so it must be *erectus*. But it was found with artifacts just like those made by modern humans. To assume the little Floresians made these artifacts contradicts almost everything that paleoanthropologists have been taught: that we didn't start to make tools until our brains were about twice the size of chimpanzees', which are approximately the size of *Homo floresiensis*. This is such a paradoxical finding!

I think the paleoanthropology community is going through the same learning process as the reviewers did. They started with the assumption that these were modern human artifacts and a pathological skull, but eventually came to accept that everything was the work of a downsized *erectus*.

Gitschier: I like it when people are forced to rethink their dogma! What about the other side of the coin—stories that you missed?

Wade: I think the main one in that category is RNAi [RNA interference], about which I've written only one story. I kept thinking, "This is fascinating, but the general reader won't be interested in the details of molecular biology, so let's wait till it advances more." I think I was far too late.

Another thing that is very difficult for science reporters to tackle is the fact that most scientific research ends nowhere. People can be very enthusiastic about what they are doing, but just as most drugs fail in clinical trials, many advances that seem very promising don't lead anywhere. So after you have been mistaken a certain number of times, you tend to be a little cautious. Of course, it's then very easy to become far more skeptical than one should be.

Gitschier: Gene therapy, for example, is a field that many thought had promise. It had some successes and some spectacular failures.

Wade: That's a field that's been going on for about 15 years. And almost all the coverage throughout the first ten years kept saying gene therapy is great. But in retrospect, it was quite wrong—it wasn't great at all. There were technical obstacles that have still not been overcome. I think the lesson for reporters is that they should not get too caught up in scientists' enthusiasm. It's fine to report that scientists are enthused about some new finding or project, but reporters should remain detached about whether or not it will succeed.

Stem cells are a case in point. The hidden premise of proposals for stem cell therapy is that we needn't understand exactly what is going on because if you just put the cells in the right place they will know what to do. My fear is that we need to understand the total cell circuitry to get stem cells to do anything useful, and that won't happen for years.

Gitschier: By choosing to write up certain stories and ignoring others, you are making judgments. Are there things out there that you are not writing about because you simply don't agree with them?

Wade: The only criteria that reporters are trained to apply is "Is this news?" So it doesn't matter if you agree with it or not.

Gitschier: But is that the ethical thing to do?

Wade: If someone makes a newsworthy claim that I suspect is not true, I will try to see if there are skeptics and expose readers to both sides of the issue. A reporter's job is to give readers sufficient information to make up their own minds. In a news story, you should expose people to all the possibilities, but you don't have to decide which one is correct. The hard thing about writing editorials is that you have to decide.

Gitschier: It's such a responsibility, I would think.

Wade: If you try to figure out the consequences of every article, you'd never write anything.

Gitschier: Returning then to the question of editorial writing, what were some of the memorable topics that you had to write about?

Wade: There weren't that many scientific issues about which we could have an editorial opinion, since many issues in science are a matter of ascertainable fact, not opinion.

I was writing editorials during the Reagan administration, so there were many environmental issues to inveigh about, inspired by the likes of Ann Gorsuch and James Watt. During the Reagan military buildup, I also wrote many editorials about military hardware and procurement scandals. I remember having great difficulty making up my mind about a "big science" project dear to physicists, the superconducting supercollider. I wrote one editorial in favor of it, the next year one against it, and the third year one in the middle. Editorial writers have to do their learning in public, or at least I did. Life is much easier once you have developed your position on the issues.

Gitschier: Today, there would seem to be a lot of opportunities for editorial writing in genetics—embryonic stem cells, cloning, reproductive choices, and intelligent design, to name a few.

Wade: I think you're right. Of course, stem cells would be less of an issue if the government hadn't tried to restrict the research. Intelligent design is a good subject for editorials, though not, I think, for the science section because it has no scientific content. It's a debate that was settled in the 19th century. It's not our role to educate people, and I see no more reason to discuss whether intelligent design is an alternative to evolution than to discuss whether or not the earth is flat.

Gitschier: What about the urgency to write things?

Wade: There is nothing like a deadline for concentrating your mind. Some of the hardest stories are when you are asked to get a story at very short notice, such as late at night when the editors see the *Washington Post* has some story, and ask you to match it. If you don't have the home numbers of the people you need to talk to, you're out of luck. Most reporters know their beat well enough that they can match a story at short notice. Fortunately, it doesn't happen too often.

Gitschier: One of the things I like about my job as a geneticist is that there is always something new on the horizon. You must feel the same way.

Wade: Yes, and journalists have the luxury of being able to move from one field to another. If it's a slow week in genetics, I can write about cognitive science.

If you're not learning something new every day, you have no one but yourself to blame.

Off the Beaten Path

An Interview with Spencer Wells

All people living today bear traces in their DNA of their ancestors. By studying the DNA of contemporary human populations, we learn how these populations relate to each other genetically, and we can make inferences into the history of how populations became dispersed through migration or war, or commingled with each other to found new ethnic groups. This interview discusses a massive effort that is underway by the National Geographic Society to tackle these questions by engaging peoples throughout the world to participate.

Interviewed October 12, 2006

Published March 30, 2007

S PENCER WELLS HAS THIS GENETICIST'S DREAM JOB. He transits the globe, collecting DNA samples, building collaborations, and orchestrating what may be the most extensive and fascinating project on human origins yet, sponsored by one of the most respected institutions in the US—The National Geographic Society. There, Wells holds the oxymoronically named post of "Explorer-in-Residence" and runs the Genographic Project (http://www.nationalgeographic.com/genographic), whose mission is to collect and genotype the Y and mitochondrial chromosomes from people the world over in order to track male and female lineages, respectively, and thereby infer migratory patterns throughout human history.

I was curious to learn how someone so young (he started working with National Geographic when he was only 33) could plunge into a project of this magnitude. The answer is that Wells is a man of many facets, vision, and energy. And he has a knack for creating the opportunity.

Happily, I had less difficulty pinning him down than I had anticipated. He suggested getting together at the American Society of Human Genetics meeting in New Orleans, where he was scheduled to give a talk. We held our interview at the open-air Café du Monde in the old French Quarter, where we were embraced by a warm breeze and the cacophony of traffic, tourists, and two busking saxophonists. Our conversation ranged over two cups of café au lait; six continents; and a medley of "The Pink Panther," "As Time Goes By," and two rounds of "Somewhere Over the Rainbow."

Picture, if you will, an Indiana Jones type, passionately delving into ancient mysteries, but in Wells's case, sunburned, hatless, and minus the whip.

Gitschier: Let's talk about your transition from academia to adventure.

Wells: Well, it was kind of roundabout. I had done my PhD at Harvard, and like many geneticists in the late '80s and early '90s, I was working on a model organism, and in this case, because my advisor was [Richard] Lewontin, it was *Drosophila*.

I did basic molecular evolution stuff, trying to detect selection in protein coding regions for an enzyme that was important for flight. I looked at variation across species and within species and did detailed statistical analysis. I found no evidence of selection, but lots of evidence of population structure. And at the end of the day I wasn't terribly interested in the population structure of fruit flies, but I had always been interested in human history.

So I wanted to apply those methods to humans, and the technology in genetics was getting to the point where you could start to study human population genetics, because back in the '80s it had been quite difficult. The human genome was so big and you had to clone everything. With PCR it got a lot easier.

Lewontin said, "You've got to work with Luca Cavalli-Sforza," out at Stanford. When I got to Stanford, one of the first things that Luca said to me was that I needed to get out into the field and meet some of the people that I was thinking of studying. Not only do we need the samples, but it's also important to hear their stories and to get to know them—to become an anthropologist, in effect.

Gitschier: Did Luca encourage everyone to do that?

Wells: He did. I had always been fascinated by Central Asia and I put together my first short expedition there for the summer of 1996. I spent five weeks in Uzbekistan, Kyrgyzstan, and Kazakhstan collecting samples.

Gitschier: Wait a second. Just hopping on a plane and going to Uzbekistan in 1996—how did you make those connections?

Wells: Good question. We knew next to nothing about Central Asia, which is why it was so fascinating. I sent off letters to the US Embassies in all of the "stans"—the newly independent stans—Kazakhstan, Turkmenistan, Tajikistan, Uzbekistan, and Kyrgyzstan. I asked the US Embassies if they could suggest any local scientists who might be interested in working with me on collecting local samples and doing a DNA study.

Most of the embassies came back and said, "No, we can't help, we don't know anybody who would be interested in this."

But the University of Tashkent, the capital of Uzbekistan, relayed it to the Academy of Sciences. And eventually, Ruslan Ruzibakiev, the director of the Institute of Immunology, happened upon it. He thought it sounded fascinating, so he wrote back to me. That was the start of a great relationship.

Gitschier: What exactly did you suggest to him?

Wells: I told him that we know very little about this very important region of the world. If you look at Eurasia, and you think about people coming out of Africa and populating the planet, this region must have played a very important role—certainly the populating of Asia, but potentially the populating of the Americas as well, and possibly India. But the very little we did know about it suggested that it was unusual. It wasn't just a mix of East and West. There might be a unique indigenous group of people who always lived there.

I told him about the new DNA markers, microsatellites, and SNPs, and this thing called the Y chromosome we were working on at Stanford and how it's starting to reveal some interesting patterns. What if we collected samples and did some studies to try to figure out how these people fit into the world pattern?

And he said, "Yeah, sounds amazing. I can set up all the details locally. You deal with all the other logistical stuff, getting yourself over here and funding it, but I know the people we need to contact."

Gitschier: How much later after you had corresponded did you take off?

Wells: He got back to me early in 1996, and by the summer, after a stack of faxes, I went.

Gitschier: How did you fund it?

Wells: That was funded with money from my Sloan Foundation post-doctoral fellowship. They give you $8,000 for reagents and supplies, and I used about $5,000 to get over there and bring all the equipment.

Gitschier: You went by yourself?

Wells: I went with a photographer, Mark Read, an English friend of my girlfriend at the time. He said, "Nobody has pictures of this part of the world. I'd love to come with you if you wouldn't mind—I have nothing to do this summer."

So I said, "Yeah, sure, sounds like fun." So the two of us set off to Central Asia knowing very little about it, speaking very little Russian.

Gitschier: Is Russian the language of all those countries?

Wells: It's the lingua franca because all those countries were part of the Soviet Union.

When we got off the plane, Ruslan picked us up with other members of the Institute. We went to his office—it was 8 o'clock in the morning—he gave us shots of vodka and said, "We have two variants: first variant is we rest today and work tomorrow, and the second variant is work today. I think we'll take the second variant."

So I was making buffers in the lab and getting ready for the first part of our expedition out of the capital two hours later.

Gitschier: Who drew the blood? This always seems like a daunting logistical problem to me.

Wells: I drew some of the blood, but we mostly worked with local phlebotomists, nurses, and so on. I was taught by one of the nurses there.

Gitschier: Did you have consent forms for all these people?

Wells: We did. We mostly took oral consent but we did have people sign in most cases. We tried very hard to explain what the project was all about.

Gitschier: This is a good point for me to ask you about something called your "blood speech."

Wells: That's a term we used in *Journey of Man* when we were filming it. [*Journey of Man* is also the title of Wells's expertly written book on the use of DNA to track human origins. His new book on the Genographic Project is called *Deep Ancestry*.]

Most people are interested in their history, and indigenous people, who are the ones who give us the clearest glimpse of their genetic history, are particularly interested, because in many cases it is all they have—what they cling on to—their sense of identity.

I was just in Tajikistan a week and a half ago, and we were sampling all over the southern part of the country and asking people to name their grandparents and great-grandparents and so on. I could do that back to maybe to my great-grandparents. These people can do it back six, seven, eight generations. They've always lived in the same place and beyond that they know even more about their history, but not necessarily their names.

So they have a sense, a clear idea of where they came from, that something is passed from generation that ties them to their ancestors. You explain to them that that thing is DNA and that it will tell us not only about the people they can name but also people beyond that that they can't name, and also people on the other side of the world—me, people you've never met in Africa or southeast Asia, and so on.

People tend to get really excited about that, I find. Generally we get a very positive response. They want to know more. They say, "I'll give you the sample, but make sure you get the information back to me, and tell me what it's all about."

Gitschier: Do you get back to them?

Wells: Yes, but not necessarily every single person individually. Sometimes we do a press interview and it will be broadcast into an area that we've worked in, and we'll talk about the group and what their DNA samples have told us.

Gitschier: So you came back to Stanford with these bloods. Had you extracted DNA as you were collecting?

Wells: We made a white cell lysate into a high SDS, high Tris buffer, so they were stable in field conditions and didn't have to be refrigerated for weeks or even months.

Gitschier: How did you get them all through customs?

Wells: It turns out there are no regulations about the import of DNA in the United States. DNA samples can be received from anywhere. They can't be sent *out* of

China or out of India, there are very strict controls, but in terms of importing DNA, if it's noninfectious, absolutely.

Gitschier: Now we're back at Stanford sometime into 1997.

Wells: And it's clear that we're seeing some interesting stuff in the samples, but we can't make a lot of sense of it because we've got this one little dot on the [geographic] map with a huge area around it that hasn't been sampled. So what we needed to do was increase the sample size from the surrounding regions.

Gitschier: And this informed your thinking in going back to Central Asia in 1998. Tell me about it.

Wells: That we should do it as a huge trip. And we should do it with a media component.

Gitschier: Why did you feel that way?

Wells: Because it's interesting! It's fun to tell people about these things. The Internet was just becoming big then. The BBC had just launched their new Web site, which was quite big in the UK, where Mark Read was based. He knew somebody there who was interested in covering the trip. It was also in part to help attract the funding for it, because a six-month overland expedition is expensive, and you need a vehicle.

So we approached Land Rover, and they were very interested in this and gave us a brand new Discovery. Virgin Atlantic agreed to fly all the equipment and us over to the UK, where we were going to start the trip, for free.

Gitschier: Just for people like me who might want to try this kind of approach, how did you pull this off?

Wells: You write a letter.

Gitschier: To whom?

Wells: To a contact. For Land Rover, it was through a family friend, who knew someone on the board. The letter gets into the right hands. In the case of Virgin Atlantic, it was just writing out of the blue to their marketing or public relations department. Just selling people on a sexy idea. The Web site, I think, did play a role in getting the funding. It provides a place for the person who gave you the equipment to be seen. And it was a new idea at the time. Not a lot of people were doing live expeditions on the Internet.

Gitschier: Does Virgin Atlantic fly to the stans?

Wells: No, we started in London and drove.

Gitschier: Oh, God. Why did you do that?

Wells: Because I liked the challenge of it.

Gitschier: Yeah, but how many thousands of extra miles is it?

Wells: It was also because I liked the idea of seeing the transition from Europe to Asia, and because we were going to start sampling in the Caucuses and the flights would have been very expensive, trying to hop between all these countries and hiring local cars and all of that. It made more sense to have a car once we got there, and it was just a question of getting it there. And it turns out it's a pretty quick drive. If you went straight through it would take you five days to get to the Georgian border.

Gitschier: OK, you are planning this trip.

Wells: I decided to leave Luca's lab to do this trip. It's hard to convince anybody, even Luca, to pay you for six months while you're out in the field. And I didn't know where I was going to end up afterwards.

Gitschier: So who actually funded it?

Wells: Luca's NIH grant did pay for the Vacutainers and the reagents.

I had applied for grants from various organizations and the anthropology funding groups would say, "You're doing genetics, you should be able to get NIH funding for this," and the NIH comes back and says, "We don't want to do human diversity or human origins research." And that's part of the reason we're funding Genographic the way we're funding it [partially through public involvement, below].

Gitschier: I know the frustration. Who else was on the trip?

Wells: Nat Pearson, who had been an undergraduate at Stanford and was starting grad school at the University of Chicago that fall. Mark Read. The journalist Darius Bazeran, who left us in Iran, and then Ruslan Ruzibakiev met us in Georgia.

Gitschier: Was the journalist posting the things on the Web?

Wells: He posted stuff for the BBC. The day we left, the BBC had a big story on the front page about "Groundbreaking US/Uzbek team leaves for Central Asia."

We did the posts [http://popgen.well.ox.ac.uk/eurasia] ourselves—I would write one, then I would edit everything, do the html, upload it, often at four in the morning before we had to leave at six a.m. Then I slept on the road. It was really fun. We had a digital camera and we posted sounds with RealAudio from people singing in cathedrals in Georgia, counting to ten in Lezgi—whatever.

Gitschier: So you went around Central Asia, giving the blood speech, collecting blood, engaging people. Six months is a long time!

Wells: Toward the end, we had two graduate students from Oxford—Matt Webster and Tatiana Zerjal—and they spent two to three weeks with us out in the field to experience it collecting samples themselves. I think they had a good time and learned a lot. That forms the core of what Tatiana did for her thesis on Genghis Khan [i.e., evidence for a common Mongol Y chromosome spread throughout Asia].

Gitschier: Was there any point during this at which you felt despair? "Oh my god why am I doing this?"

Wells: Absolutely. A lot of fieldwork is incredibly boring. A lot of time is spent waiting for permissions, sitting in meetings, waiting for people to show up. Or you blow into town on a Saturday afternoon in one of the ugliest places in the world and you have a day and half to waste.

Gitschier: So what did you do?

Wells: You read or talked. But I did question what the hell I was doing out here. I didn't have a job lined up for when I returned.

Later, a possibility presented itself to work with Walter Bodmer at Oxford. So the samples moved to Walter's lab. And post-docs started coming through.

Then things started to get really exciting scientifically. We had a huge number of samples. In addition to the ones we collected on the trips, our collaborators continued to collect, so we ended up with over 2,000 samples and amazing Y chromosome markers that Peter Underhill and Peter Oefner discovered. The combination was incredible. Every experiment we ran was exciting and new. You're getting these results and they start to make sense. Piecing together migratory patterns. As we started to publish papers, the popular press picks up on it because the Y chromosome is such a great tool for telling these stories and because it actually reveals migratory routes and ties in with historical figures, in the case of Genghis Khan and so on.

Eventually I was contacted by a film production company in London that was working on a film about human variation. Are we all the same, or are we different—the whole race issue. They interviewed me and the producer said, "This stuff is absolutely amazing," and suggested, "We should do a film just on what you guys are doing," and that became the *Journey of Man*.

Then, people at National Geographic wanted to have a meeting with me because *Journey of Man* was going to be one of their big television specials. That was the genesis of the Genographic Project. We developed it over the course of the next couple of years, organized the funding, and it is now launched.

Gitschier: For our readership—the goal of the Genographic Project is to...

Wells: To use DNA as a tool to answer that basic human question—where do we all come from.

Gitschier: And the way you are going to do that is...

Wells: Is by studying genetic markers from people from around the world, focusing particularly on indigenous groups because they retain that geographic context in which the genetic patterns originated to a greater extent than people like me—I've got ancestors from all over northern Europe and I live on the eastern seaboard of north America.

Gitschier: And you're doing this via collaborations with people all over the world.

Wells: We have ten regional centers focusing on sampling extant populations. One center is devoted to ancient DNA, which is a very important component of it.

Gitschier: Does National Geographic support these people directly?

Wells: Yes, we essentially give them multi-year grants. They are contracts, and there are deliverables.

Gitschier: So these centers do the fieldwork, make the DNA, and do the genotyping, and then?

Wells: The data are sent to a central database created by IBM. They supplied the server, which is sitting in the basement of National Geographic. They have given everybody laptops with biometric [i.e., fingerprint] recognition so that only the PIs can access to the database. We are working closely with their computational biology team on analyzing the data. So some of our first publications, which are starting to go into the journals now, are coming through that group.

Gitschier: Are you at liberty to say what these are about?

Wells: We are expanding the survey of whole mitochondria genomes in Africa. We've doubled the size of that database and it's revealing interesting mitochondrial patterns. That effort has been spearheaded by Doron Behar in Haifa.

People tend to ignore what went on *within* Africa. There is this inherent bias in European and Asian scientists that we've "done" Africa and *then* things got interesting when we [humans] left, but of course there was still a lot going on within Africa. We're looking at routes people might have taken out of Africa and back migration into Africa. Information that is coming out, in part, from an expedition I organized in 2005 to the Tibesti mountains in Chad, up on the Libyan border.

Gitschier: There is also the public participation component of the project. How many people have sent in their 100 bucks and their cheek swab [for DNA analysis]?

Wells: Around 165,000. It's been a great response.

Gitschier: What is the reaction from people when they get their ancestral information?

Wells: It runs the gamut, from people who write to tell us how amazing the project is and how they've learned what it is to be human, to people who say "I already knew I was Western European, you didn't tell me anything new." But mostly people are very positive about it. I think it is really tapping into something, certainly in North America—this desire to learn more about where we came from. We are a nation of immigrants, so it's not too surprising.

We've gotten some fascinating results and a lot of e-mails. For example, a Hungarian woman wrote in and said, "You've got to redo my test. You told me I'm native American or Siberian, and I know my ancestors came from Hungary—I can tell you the village they were living in in the sixteenth century." The Hungarian language, Magyar, is actually related to languages spoken in Siberia, and this is one of the first cases where we've actually seen Siberian lineages showing up in the Hungarian population. They are there at very low frequency. We now through this project have over 350 people who are of Hungarian descent and we see these [Siberian] lineages at four to five percent on both male and female sides.

Gitschier: What kind of information do people give you when they sign on to this project?

Wells: It is totally anonymous. We don't actually know which number goes into which kit. You get a randomly generated alpha-numeric code and that is the only way you can access your results on the Web site. When you log on, the first thing that pops up is a sign that says, "please help us with the project by telling us more about yourself and donating your information to the database." So we ask people their sex, their birth date, zip code, language of their parents, origin of their earliest known male ancestor, and so on. So it provides a little bit of anthropological context.

Gitschier: I'd like to do it myself.

Wells: Well, we can send you a kit.

Public participation has raised around sixteen million gross. The net proceeds are around four million. This is a nonprofit endeavor, and we're plowing that money back into the project, with half going to the field research and the other half going to the "legacy" fund to fund projects by indigenous groups. For example, a project on language preservation, or to fund a Middle Eastern women's cooperative to resurrect a particular type of traditional embroidery work. We're giving our first grants before the end of the year. They are typically around $25,000 each. It's a way to give something back to the indigenous people whose way of life is endangered in some places.

Gitschier: Shifting gears, I'm interested in the general question of how people end up doing what they're doing. As a child, what was your thinking?

Wells: I wanted to be a historian or a writer when I was very young.

Gitschier: Why?

Wells: I was fascinated by history. It was just this idea that there were people who lived in a different way in a different time. And you could imagine yourself, through reading a book, being there with them—like time travel.

I wasn't that interested in science as a kid. I collected rocks and insects just like any other kid, but in terms of a career, I wasn't that excited till my mother went back to graduate school to get her PhD in Biology and I started hanging out in the lab with her and discovered that science was really fun and cool. It's not just about geeky guys in white lab coats, it's about solving puzzles on a daily basis.

I decided I wanted to combine the two in some way. I studied molecular biology in college, at the University of Texas at Austin.

My father is a lawyer, a tax attorney. I was never tempted to follow his footsteps. But he comes from a military family. His father, whom he never knew because he died in World War II, graduated top of his class from West Point and was apparently a wild man out in the field. I think maybe I got some of my love of danger and going to strange places from him.

Gitschier: Give me an example or two.

Wells: Everything is interesting in its own way. Spending time with the Chukchi reindeer herders, where one morning in Siberia it got down to $-70°C$. I had never experienced cold like that—dry ice levels almost. It takes you out of yourself. You think that life is all about certain things, but something like that causes you to sit up and think there are some pretty extreme, amazing things in the world. And the fact that we were there filming people who were living in a traditional way in animal skins and we were in thick down coats—that *really* gives you pause! People have lived like this for tens of thousands of years.

I'm very, very lucky that I have the opportunities that I do, to travel around the world and meet people and hear their stories and see their way of life. It makes me very sad to see that those ways of life are dying.

Gitschier: The Genographic Project is a model for doing things outside of a university setting, for getting the public actively involved in a real-time scientific enterprise, and to help fund it. I think it's fantastic.

Wells: The only advice I would give to young scientists is to think outside the box. Don't just do what your advisor did. Don't just do what the other graduate students are doing. If you have interests that seem a little bit flaky, off the beaten path, that's probably a good thing—especially if they are passions.

Further Reading

Victor Ambros

Lee RC, Feinbaum RL, Ambros V. 1993. The *C. elegans* heterochronic gene *lin-4* encodes small RNAs with antisense complementarity to *lin-14*. *Cell* **75:** 843–854.

Lee RC, Ambros V. 2001. An extensive class of small RNAs in *Caenorhabditis elegans*. *Science* **294:** 862–864.

Adrian Bird

Bird AP, Southern EM. 1978. Use of restriction enzymes to study eukaryotic DNA methylation. I. The methylation pattern in ribosomal DNA from *Xenopus laevis*. *J Mol Biol* **118:** 27–47.

Bird AP. 1986. CpG-rich islands and the function of DNA methylation. *Nature* **321:** 209–213.

Guy J, Gan J, Selfridge J, Cobb S, Bird A. 2007. Reversal of neurological defects in a mouse model of Rett syndrome. *Science* **315:** 1143–1147.

David Botstein

Botstein D, White RL, Skolnick M, Davis RW. 1980. Construction of a genetic linkage map in man using restriction fragment length polymorphisms. *Am J Hum Genet* **32:** 314–331.

Herb Boyer

Boyer H. 1964. Genetic control of restriction and modification in *Escherichia coli*. *J Bacteriol* **88:** 1652–1660.

Cohen SN, Chang ACY, Boyer HW, Helling RB. 1973. Construction of biologically functional bacterial plasmids in vitro. *Proc Natl Acad Sci* **70:** 3240–3244.

Pat Brown

Schena M, Shalon D, Davis RW, Brown PO. 1995. Quantitative monitoring of gene expression patterns with a complementary DNA microarray. *Science* **270:** 467–470.

DeRisi JL, Iyer VR, Brown PO. 1997. Exploring the metabolic and genetic control of gene expression on a genomic scale. *Science* **278:** 680–686.

Rebecca Cann

Cann RL, Stoneking M, Wilson AC. 1987. Mitochondrial DNA and human evolution. *Nature* **325:** 31–36.

Sean Carroll

Carroll SB, Scott MP. 1985. Localization of the *fushi tarazu* protein during *Drosophila* embryogenesis. *Cell* **43:** 47–57.

Carroll SB, Gates J, Keys DN, Paddock SW, Panganiban GEF, Selegue JE, Williams JA. 1994. Pattern formation and eyespot determination in butterfly wings. *Science* **265:** 109–114.

Carroll S. 2005. *Endless forms most beautiful: The new science of evo devo.* Norton, New York.

Tom Cech

Cech TR, Zaug AJ, Grabowski PF. 1981. In vitro splicing of the ribosomal RNA precursor of *Tetrahymena*: Involvement of a guanosine nucleotide in the excision of the intervening sequence. *Cell* **27:** 487–496.

Kruger K, Grabowski PJ, Zaug AJ, Sands J, Gottschling DE, Cech TR. 1982. Self-splicing RNA: Autoexcision and autocyclization of the ribosomal RNA intervening sequence of *Tetrahymena*. *Cell* **31:** 147–157.

Soraya de Chadarevian

de Chadarevian S. 2002. *Designs for life: Molecular biology after World War II.* Cambridge University Press, Cambridge.

Evan Eichler

Bailey JA, Gu A, Clark RA, Reinert K, Samonte RV, Schwartz S, Adams MD, Myers EW, Li PW, Eichler EE. 2002. Recent segmental duplications in the human genome. *Science* **297:** 1003–1007.

Sharp AJ, Hansen S, Selzer RR, Cheng Z, Regan R, Hurst JA, Stewart H, Price SM, Blair E, Hennekam RC, et al. 2006. Discovery of previously unidentified genomic disorders from the duplication architecture of the human genome. *Nat Genet* **38:** 1038–1042.

Jenny Graves

Sinclair AH, Foster JW, Spencer JA, Page DC, Palmer M, Goodfellow PN, Graves JAM. 1988. Sequence homologous to ZFY, a candidate human sex-determining gene, are autosomal in marsupials. *Nature* **336:** 780–783.

Warren WC, Hillier LW, Marshall Graves JA, Birney E, Ponting CP, Grützner F, Belov K, Miller W, Clarke L, Chinwalla AT, et al. 2008. Genome analysis of the platypus reveals unique signatures of evolution. *Nature* **453:** 175–183.

Sir Alec Jeffreys

Jeffreys AJ, Wilson V, Thein SL. 1985. Hypervariable 'minisatellite' regions in human DNA. *Nature* **314:** 67–73.

Jeffreys AJ, Brookfield JFY, Semeonoff R. 1985. Positive identification of an immigration test-case using human DNA fingerprints. *Nature* **317:** 818–819.

Judge John E. Jones III

Kitzmiller et al. v. Dover Area School District et al. 2005. Case No. 04cv2688. Memorandum Opinion.

Mary Lyon

Lyon MF. 1961. Gene action in the X-chromosome of the mouse (*Mus musculus* L.). *Nature* **190:** 372–373.

Lyon MF. 1961. Sex chromatin and gene action in the mammalian X-chromosome. *Am J Hum Genet* **14:** 135–148.

Svante Pääbo

Gree RE, Krause J, Ptak SE, Briggs A, Ronan MT, Simons JF, Du L, Egholm M, Rothberg JM, Paunovic M, Pääbo S. 2006. Analysis of one million base pairs of Neanderthal DNA. *Nature* **444:** 330–336.

Pääbo S. 1985. Molecular cloning of ancient Egyptian mummy DNA. *Nature* **314:** 644–645.

Neil Risch

Burchard EG, Ziv E, Coyle N, Gomez SL, Tang H, Karter AJ, Mountain JL, Pérez-Stable EJ, Sheppard D, Risch N. 2003. The importance of race and ethnic background in biomedical research and clinical practice. *N Engl J Med* **348:** 1170–1175.

Elaine Strass

Not applicable.

Sir John Sulston

Sulston JE, Horvitz HR. 1977. Post-embryonic cell lineages of the nematode *Caenorhabditis elegans*. *Dev Biol* **56:** 110–156.

Sulston JE, Schierenberg E, White JG, Thomson JN. 1983. The embryonic cell lineage of the nematode *Caenorhabditis elegans*. *Dev Biol* **100:** 64–119.

Sulston J, Ferry G. 2002. *The common thread: A story of science, politics, ethics, and the human genome.* Joseph Henry (National Academies Imprint), Washington, D.C.

Jamie Thomson

Thomson JA, Itskovitz-Eldor J, Shapiro SS, Waknitz MA, Swiergiel JJ, Marshall VS, Jones JM. 1998. Embryonic stem cell lines derived from human blastocysts. *Science* **282:** 1145–1147.

Yu J, Vodyanik MA, Smuga-Otto K, Antosiewicz-Bourget J, Frane JL, Tian S, Nie J, Jonsdottir GA, Ruotti V, Stewart R, et al. 2007. Induced pluripotent stem cell lines derived from human somatic cells. *Science* **318:** 1917–1920.

Shirley Tilghman

Tilghman SM, Tiemeier DC, Seidman JG, Peterlin BM, Sullivan M, Maizel JV, Leder P. 1978. Intervening sequence of DNA identified in the structural portion of a mouse β-globin gene. *Proc Natl Acad Sci* **75:** 725–729.

Bartolomei MS, Zemel S, Tilghman SM. 1991. Parental imprinting of the mouse H19 gene. *Nature* **351:** 153–155.

Nicholas Wade

Refer to *The New York Times*.

Spencer Wells

Wells S. 2004. *The journey of man: A genetic odyssey*. Random House, New York.

Wells S. 2007. *Deep ancestry: Inside the Genographic Project*. National Geographic Society, Washington, D.C.

Index

Abbott, Cathy, 12
Abelson, John, 78
Adams, Mark, 101
Adelberg, Ed, 35, 36
Affymetrix, 48, 102
Agilent, 102
Aldrich, Judy, 38
Alpha-fetoprotein (AFP), 217–218
Ambros, Victor
   anti-sense base pairing
      determination, 7
   collaboration with Ruvkun's lab, 7–8
   education, 2
   family background, 3–4
   inspiration for a career in
      science, 2–3
   introductory information, 1
   lin-4 research beginnings, 5–7
   Ras pathway discovery, 4–5
   RNAi association with *lin-4*, 8–9
   on winning the Lasker Award, 10
American Civil Liberties Union
   (ACLU), 128
American College of Medical
   Genetics, 177
American Society for Gene Therapy, 177
American Society of Human Genetics
   (ASHG), 171, 172, 173, 174–175
Amersham, 15
Anderson, Dorothy, 27
Animal farming elimination project, 51–52
Antennapedia complex, 65
Anticipation phenomenon, 96
Arber, Werner, 36
Argonaut, 8
ASHG. *See* American Society of Human
   Genetics
Ashkenazi Jews, 165–166
Autism, 19, 116
Axolotl, 4

Bailey, Jeff, 99
Baltimore, David, 10, 76, 216

Barber, John, 97
Bartolomei, Marisa, 218
Baulcombe, David, 8, 9
Baumann, Peter, 79
Baylor College of Medicine, 94, 95
Bazeran, Darius, 237
BBC, 237
Behar, Doron, 239
Behe, Michael, 135
Bell, Graeme, 121
Bender, Welcome, 65
Berg, Paul, 38, 40
Bialek, Bill, 26
Bird, Adrian
   coining of the name "CpG
      islands," 15
   determination of a MeCP2 association
      with neuromaintenance
      disorders, 19
   discovery of MeCP2 association with
      Rett Syndrome, 16–18
   exploration of function of
      MeCP2, 18–19
   possibility that Rett Syndrome is
      reversible, 20
   purification of MeCP2, 16
   result of MeCP2 knockout experiment,
      16–17
   span of work, 12
   start of interest in methylation,
      13–15
   start of work on protein binding to
      methylated regions, 15–16
Birnstiel, Max, 13, 15
Bishop, Jerry, 28, 38
Blackburn, Liz, 78
Blaxter, Kenneth, 143
Bloch, Konrad, 23–24
Bodleian library, 90
Bodmer, Walter, 144, 238
Bonobo ape, 159
Borst, Piet, 117, 118
Botstein, Ania Wyszewianska, 27

Botstein, David
　family background, 27–28
　genesis of mapping paper, 28–30
　graduate curriculum design, 25–26
　introductory information, 22
　music background, 26–27
　start at Lewis-Sigler Institute, 22–23
　undergraduate curriculum design, 24–25
　on why students don't become professional scientists, 23–25
Bouck, Noel, 36
Boyer, Herb
　early experiments with bacterial conjugation, 35–36
　EcoRI identification, 37–38
　events surrounding the cleaving of pSC101 with EcoRI, 39–40
　family background, 32–33
　introductory information, 32
　mapping of alleles of K12 and B, 36
　on not winning a Nobel Prize, 40–41
　search for recognition site, 36–37
　start of restriction and modification work, 33–35
Bradley, Allan, 192
Brenner, Sydney, 182
Bressman, Susan, 166
Brown, Don, 13
Brown, Gene, 24
Brown, Pat
　animal farming elimination project plans, 51–52
　background to start of microarray work, 44–46
　capillary printing use, 47
　events leading to expression experiment publication, 48–50
　family background, 44
　introductory information, 44
　PLoS Genetics and, 51
　robotics used for DNA gridding, 47–48
　tissue staining grant application, 50
　work with Dari Shalon, 46–47
Brown, Wes, 57–58
Bryan, William Jennings, 130
Buffett, Warren, 196

Cambridge University, 84
Cann, Rebecca
　application of mitochondrial analysis to human evolution, 58–59
　conclusions from mitochondrial DNA research, 54
　early genetics work, 55–56, 59
　family background, 54–55
　gender bias experience, 61
　mitochondrial fractions work of Wes Brown, 57–58
　PAUP algorithms' impact on research direction, 59–62
　reactions to her work, 62
　start of interest in restriction enzymes, 56
Carroll, Sean
　early fly embryo work, 65–66
　experimental work on butterflies, 69
　growing interest in evolution, 64–65
　interest in paleontology, 70–71
　introductory information, 64
　move to butterflies to study evo devo, 68–69
　success in localizing *ftz* protein, 66–67
　on teaching evolutionary science through books, 73
　writing career, 68, 70, 71–73
Carter, Toby, 144, 145
Case Western Reserve University, 98
Caskey, Tom, 95
Cavalli-Sforza, Luca, 60, 62, 233, 237
Cech, Tom
　background to RNA splicing work, 77–79
　on the discovery of RNA as a catalyst, 79
　early work on psoralen cross-linking, 76–77
　future prospects, 82
　introductory information, 76
　new interest in neurosciences, 81
　on scientists writing about their work, 80
　on taking the job at HHMI, 80–81
C. elegans
　embryonic imaging documentation, 184–187
　introductory information, 181, 182
　nematode genome studies, 190–192
　Sulston's work on (*see* Sulston, John)
　technique used to study live worms, 183–184
　technology improvements for studying embryos, 187

Celera, 100, 226
Chakravarti, Aravinda, 98
Chambon, Pierre, 120
Chan, Russell, 29
Chang, Annie, 40
Charo, Alto, 205
Chovnick, Art, 173
Cohen, Stan, 32, 33, 39
Cold War, 88
Collins, Francis, 101
Combinatorial screen, 207
Common Thread, The (Sulston), 182, 193
Cooper, David, 14
Cooper, Des, 107
Copy number variation research, 97–98, 116–117
Coulson, Alan, 191
Cowan, Max, 81
CpG islands, 12, 15
CPTech, 196
Creationism. *See* Intelligent design
Crick, Francis, 88, 91, 92
Curt Stern Award, 176

D14S1, 30
Darrow, Clarence, 130
Darwin, Charles, 135
Davis, Ron, 28, 29, 38
Dawid, Igor, 13
de Chadarevian, Soraya
  definition of history, 88–89
  educational background, 84–85
  e-mail's presence and impact on science collaboration, 89
  evolution of her book, 87–88
  on "experimenter's regress," 86
  introductory information, 84
  on the mystic of double helix discovery, 88, 91–92
  on the new approaches to studying the history of science, 86–87
  start of biogenetics research, 85–86
  on thinking of the audience when writing history, 91
  on the use of archives in research, 90–91
  on the use of interviews as historical material, 89–90
  on the utility of having a science degree, 85

de Crombrugghe, Benoit, 15
Deep Ancestry (Wells), 235
Delbruck, Max, 26
DeRisi, Joe, 48, 49
Designs for Life (de Chadarevian), 84
Diamond, Jared, 201
Dickey amendment, 205
DiNardo, Steve, 67
distaless, 69
Distance trees, 59, 60
DNA, recombinant. *See* Recombinant DNA
DNA chips. *See* Microarrays
DNA fingerprinting
  applied to forensics, 123–124
  applied to profiling, 124–125
  introductory information, 115, 116
  Jeffreys' work on (*see* Jeffreys, Alec)
  RFLP discovery, 120–121
  start of use of, 122–123
DNA methylation
  Bird's interest in (*see* Bird, Adrian)
  discovery of MeCP2 association with Rett Syndrome, 17–18
  introductory information, 11, 12
  work on protein binding to methylated regions, 15–16
DNA polymorphisms
  genesis of Botstein's mapping paper, 28–30
  introductory information, 21
Dolly, 200, 206
Double helix discovery, 88, 91–92
Double-stranded RNA (dsRNA), 8
Dover school district case. *See* Kitzmiller et al. v. Dover Area School District
Drosophila melanogaster, 63, 64. *See also* Evolutionary development
Dubois, Eugené, 73
Dunham, Maitreya, 26
Duplications research, 97–98, 100–102
Durant, Barbara, 201
Dussoix, Daisy, 36
Dystonia, 165–166

E912, 6
EcoRI identification, 37–38, 39–40
Edwards, Anthony, 144
Edwards v. Aguillard, 132
Effect prong in law, 137
Ehrlich, Paul, 201

Eichler, Evan
    challenges from jerry-rigging computer hardware, 99–100
    controversy over genome collaboration efforts, 100–101
    duplications and copy number variation research, 97–98
    duplications map creation, 100
    duplications research applied to evolutionary history, 101–102
    early interest in genetics, 94–95
    educational background, 95–96
    favorite research topic, 102–103
    Fragile X syndrome research, 96–97
    introductory information, 94
    move to Case Western and genome-wide studies, 98–100
Eisen, Mike, 51
E-mail and collaboration in science, 89
Embryonic stem cells
    ethical issues, 205
    future of human research, 209
    introductory information, 199, 200
    move from primate to human embryos, 204–205
    requirement to use existing cell lines, 205–206
    stem cell therapy, 227
    success with a new medium for embryos, 204
    Thomson's work on (see Thomson, Jamie)
    work on deriving primary embryonic stem cells, 202–203
    work on deriving primate embryonic stem cells, 202, 203–204
Endless Forms Most Beautiful (Carroll), 70
Endorsement test, 137, 139
Engelsberg, Ellis, 34
engrailed, 67
Epperson v. Arkansas, 131, 132, 138
Establishment Clause, 137
Eutherian mammals, 105
Evolution
    application of mitochondrial analysis to human evolution, 54, 58–59
    Carroll's interest in, 64–65
    duplications research applied to evolutionary history, 101–102
    intelligent design versus (see Intelligent design)
    legal cases challenging bans on teaching, 131
    scientist's views on intelligent design, 110–111, 228
    teaching evolutionary science through books, 73
Evolutionary anthropology. *See also* Evolutionary development; Population genetics
    contamination issues in sample collection, 156–157, 158
    development of technology to use for ancient DNA, 158
    DNA sampling use in paleontology work, 157
    extraction of DNA samples from mummies, 155–156
    Pääbo's work on (*see* Pääbo, Svante)
    start of a German institute in, 158–159
Evolutionary development (evo devo). *See also* Genomic architecture; Population genetics
    butterflies used to study, 68–69
    Carroll's work on (*see* Carroll, Sean)
    duplications research applied to evolutionary history, 101–102
    introductory information, 64
"Experimenter's regress," 86

Falconer, Douglas, 144, 145, 146
Falkow, Stanley, 39
Fanconi, Guido, 27
Feinbaum, Rhonda, 6–7
Ferguson, Chip, 5
Ferry, Georgina, 193
Finnegan, David, 117
Fire, Andy, 8, 176
First Amendment, 137
Fisher, Elizabeth, 142
Fisher, R.A., 144, 145
Fitch-Margoliash trees, 59
Flavell, Dick, 118
Fost, Norm, 205
Fox, Maury, 22, 30
Fox Chase Cancer Center, 216
Fragile X syndrome research, 96–97
Francke, Uta, 16, 17, 45
Freed, Lenny, 54
Frommer, Marianne, 15
Fu, Ying-Hui, 96
fushi tarazu (*ftz*), 64, 66–67

Gall, Joe, 13
Gardner, David, 204
Gates Foundation, 195, 196
Genentech, 32, 41
Gene therapy, 227
Genetic markers, 21, 28–30
Genetics, 59, 178
Genetics Society of America (GSA)
 awards given by, 176
 duties of the executive director, 175
 introductory information, 171, 172
 meetings run by, 173–174
 meeting structure, 177–178
 structure of GSA vs. ASHG, 174–175
Genographic Project, 232, 239, 242
Genome, 28
Genomic architecture. *See also* Evolutionary development; Population genetics
 duplications and copy number variation research, 97–98
 duplications map creation, 100
 duplications research applied to evolutionary history, 101–102
 Eichler's work on (see Eichler, Evan)
 Fragile X syndrome research, 96–97
 introductory information, 93, 94
George W. Beadle award, 176
Geron Corporation, 205
Gesteland, Ray, 65
Globin genes, 116. *See also* DNA fingerprinting
Goodfellow, Peter, 111
Goodman, Howard, 38
Gorsuch, Ann, 228
Gould, Stephen Jay, 4
Grabowski, Paula, 78
Graves, Jenny
 focus on X-inactivation in marsupials, 106–107
 on her past illness and funding issues, 112–113
 interest in science education, 110–111
 introductory information, 106
 platypus endangered status, 108
 platypus sex chromosomes studies, 109
 realization that out-group studies could help with genome mapping, 113
 on reports of monotreme fossils outside Australia, 107–108

 testes-determining factor/gene cloning story, 111
 views of intelligent design, 110
Greenberg, Mike, 18
Greider, Carol, 78
Gruneberg, Hans, 148
GSA. *See* Genetics Society of America
GSA Medal, 176
Gurvich, Gerry, 172, 173
Gusella, Jim, 30
Guy, Jacky, 20

Haemophilus parainfluenzae II (Hpall), 13
Hafen, Ernst, 66
Haussler, David, 100
Hearn, John, 202, 203
Hearst, John, 77
Helling, Bob, 40
Hemochromatosis, 29
Herskowitz, Ira, 24
Heterochronic mutants, 4–5
HHMI (Howard Hughes Medical Institute), 217
Higgs, Doug, 121
Hiroshima, 88
History of science
 definition of history by de Chadarevian, 88–89
 evolution of de Chadarevian's book on molecular biology, 87–88
 mystic of double helix discovery and, 88, 91–92
 new approaches to studying, 86–87
 thinking of the audience when writing history, 91
 use of archives in research, 90–91
 use of interviews as historical material, 89–90
 utility of having a science degree when writing, 85
HLA (human leukocyte antigen), 29
Hodgkin, Jonathan, 191
Hogness, David, 117
Hominid paleontology, 71
Homo floresiensis, 226
Homo neanderthalensi, 58, 153, 154, 157, 159, 161
Horvitz, Bob
 Ambros's work with, 4–5, 10
 GSA medal and, 176
 Sulston's work with, 183, 184, 188, 191

Howard Hughes Medical Institute (HHMI), 76, 80–81, 217
Hpall (*Haemophilus parainfluenzae II*), 13
HTF islands, 15
Hubbard, Tim, 196
HUGO (Human Genome Organization), 177
Human Genetics Commission (UK), 195
Human Genome Organization (HUGO), 177
Human leukocyte antigen (HLA), 29

Imprinting
  introductory information, 211
  realization that *H19* is imprinted, 218
  Tilghman's work on (*see* Tilghman, Shirley)
Induced pluripotent stem (IPS) cells, 200, 206–208
Intelligent design
  court case on (*see also* Kitzmiller et al. v. Dover Area School District)
  evolution versus (*see* Evolution)
  introductory information, 127
  question of whether it is a science, 136, 138
  scientist's views on, 110, 228
International Society for Stem Cell Research (ISSCR), 208
"Into the Jungle" (Carroll), 72
Introns, 120
IPS (induced pluripotent stem) cells, 200, 206–208

Jacob, François, 79
Jaenisch, Rudi, 216
Jeffreys, Alec
  copy number variation work, 116–117
  DNA fingerprinting applied to forensics, 123–124
  DNA fingerprinting story start, 122–123
  DNA profiling, 124–125
  introductory information, 116
  on kids and science, 125
  post-doc period leading to EcoRI site discovery, 117–120
  research on variation in restriction sites among people, 121
  search for variable microsatellites, 121–122

Johnston, Mark, 178
Jones, Beth, 22
Jones, James F., 68
Jones, Jeff, 204
Jones, John E. III
  background, 129
  court case concerning intelligent design (*see* Kitzmiller et al. v. Dover Area School District)
  introductory information, 128
  on the personal impact of the Dover case, 139–140
  personal view of creationism versus evolution, 134
  unfamiliarity with intelligent design, 133
  on why he doesn't engage in independent research, 133–134
Journey of Man (Wells), 235, 238
"Judgment Day," 139

K12 and B alleles, 36–37
Kaiser, Dale, 39
Kaiser Permanente, 168–169
Kallioniemi, Anne, 45
Kangaroos, 106
Karr, Tim, 66
Kaufman, Thom, 65
Kendrew, John, 87, 90
Kirschstein, Ruth, 30
Kitzmiller, Tammy, 128
Kitzmiller et al. v. Dover Area School District
  basis of case, 128
  expert opinion process, 133–134
  impact on school boards outside Pennsylvania, 138–139
  legal cases challenging bans on teaching evolution, 131
  legal structure of the case, 129–130
  performance of witnesses during the case, 135
  presiding judge (*see* Jones, John E. III)
  question of whether intelligent design is a science, 136, 138
  relevant legal definitions, 137
  states' response to Supreme Court decision, 132
Kleckner, Nancy, 29
Klug, Aaron, 197
Kovacs, Greg, 46
Krasnow, Mark, 50
Kravitz, Kerry, 29
Kruglyak, Leonid, 25

Laboratory of Molecular Biology (LMB), 188, 197
Lalande, Marc, 97
Lander, Eric, 99, 101
L-arabinose metabolic pathway, 34
Lasker Award, 9, 10
Laughon, Allen, 65
Leder, Phil, 78, 119, 219
Lee, Rosalind "Candy," 6–7, 9
Leipzig Institute, 158–159
Lemon v. Kurtzman (Lemon test), 137, 139
let-7, 9
Levine, Arnie, 216
Levine, Mike, 70
Levinthal, Cyrus, 23
Lewis, Ed, 65
Lewis-Sigler Institute for Integrative Genomics, 22
Lewontin, Richard, 232, 233
lin-14, 5
lin-4
 determination of an association with RNAi, 8–9
 research beginnings, 5–7
LINE elements, 148–150
Lingner, Joachim, 78
Linn, Stu, 37
LMB (Laboratory of Molecular Biology), 188, 197
Longitude (Sobel), 91
Loutit, John, 146
Love, Jamie, 196
Lupski, Jim, 102
Luria, Salvador, 23, 24, 26
Lyon, Mary
 criticism received from Gruneberg, 148
 early interest in science, 142–143
 expansion of educational opportunities for women during WWII, 143–144
 introductory information, 142
 mottled mice experiments, 146–147
 origin of hypothesis about LINE elements, 148–150
 post-doc work on mutagenesis, 145–146
 reflections on her career, 150
 Tabby mice experiments, 148
 university education, 143
 work with Fisher, 144–145

Macaca nigra, 202
Magasanik, Boris, 22, 23, 24
Maniatis, Tom, 119
Marsupials, 105
Max, Edward, 15
MeCP2
 association with neuromaintenance disorders, 19
 discovery of association with Rett Syndrome, 17–18
 exploration of function, 18–19
 introductory information, 12
 possibility that Rett Syndrome is reversible, 20
 purification of, 16
 result of knockout experiment, 16
Medecins sans Frontieres, 195
Meins, Frederick, 201
Melampus bidentatus, 201
Mello, Craig, 8
Mertz, Janet, 38
Meselson, Matt, 36
Microarrays (DNA chips)
 Brown's work on (*see* Brown, Pat)
 capillary printing and, 47
 introductory information, 43
 robotics used for DNA gridding, 47–48
Migeon, Barbara, 15
Miller, Ken, 134–135
miRNAs (microRNAs). *See* RNAi
Mitochondrial DNA
 application to human evolution, 54, 58–59
 Cann's work on (*see* Cann, Rebecca)
 DNA sampling use in paleontology work, 157
 introductory information, 53
 mitochondrial fractions work of Wes Brown, 57–58
 PAUP algorithms' impact on research direction, 59–62
Mitochondrial Eve, 54
Model Organism to Human Biology (MOHB), 178
Molecular biology history, 87–88
Monod, Jacques, 79
Monotremes
 Graves's work on (*see* Graves, Jenny)
 introductory information, 105, 106
 realization that out-group studies could help with genome mapping, 113
Mottled gene, 146–147

Nagasaki, 88
Nathans, Dan, 38

Index

National Geographic Society, 231, 232, 238, 239
National Society of Genetic Counselors (NSGC), 177
Nature, 61, 70, 223
Neanderthals. See *Homo neanderthalensis*
Neglected Diseases Initiative, 195
Nei, Masatoshi, 60
Nelson, David, 94, 102
Nelson, Stan, 45
Neomorphic activity, 6
Nicholls, Rob, 98
Nijhout, Fred, 68
Noller, Harry, 79
Nomarskik microscopes, 183
NSGC (National Society of Genetic Counselors), 177
Nüsslein-Volhard, Christiane, 65

O'Brien, Steve, 113
O'Connor, Sandra Day, 137
Oefner, Peter, 238
O'Farrell, Pat, 67
Ohno, Susumo, 147
Olson, Maynard, 192
Oregon Primate Center, 202
Orgel, Leslie, 193
Owen, George, 145
Oxfam, 195

P1 phage transduction, 35
Pääbo, Svante
  contamination issues in sample collection, 156–157, 158
  development of technology to use for ancient DNA, 158
  extraction of DNA samples from mummies, 155–156
  group project discussions appreciation, 159–160
  interest in ape comparative genomics, 159
  interest in further research on Neanderthals, 161
  introductory information, 154
  population history interest, 160–161
  sample gathering process, 157
  start of a German institute in evolutionary anthropology, 158–159
  start of interest in Egyptology, 154–155
Pachnis, Vassilis, 218
Page, David, 111
Pallid mutant, 145
Pardue, Mary Lou, 76
PAUP (phylogenetic analysis using parsimony), 59–62
Pearson, Nat, 237
Pepper Hamilton, 132–133
Perutz, Max, 90, 197
Phillips, Rita, 146
Phylogenetic analysis using parsimony (PAUP), 59–62
Platypus, 108, 109
PLoS Genetics, 22, 44, 51, 116, 142, 196, 224
Population genetics. See also Evolutionary anthropology
  contamination issues in sample collection, 156–157, 158
  development of technology to use for ancient DNA, 158
  extraction of DNA samples from mummies, 155–156
  Genographic Project, 239
  introductory information, 231
  Pääbo's work on (see Pääbo, Svante)
  politicizing of the human genome, 167–168, 193–195
  population blood sample collection in Central Asia, 233–236
  Risch's studies at Kaiser, 168–169
  Risch's study of dystonia in Jews, 165–166
  sample gathering process, 157
  Wells's work on (see Wells, Spencer)
Princeton University, 212. See also Tilghman, Shirley
Pritchard, Bob, 120
Protection of telomeres (POT1), 79
pSC101, 39–40
Psoralen cross-linking, 76–77
Purpose prong in law, 137

Race and genetics
  definition of race, 167
  introductory information, 163
  politicizing of the human genome, 167–168, 193–195
  population genetics study at Kaiser, 168–169
  study of dystonia in Jews, 165–166
Ras pathway, 4–5
Read, Mark, 234, 235, 237
Recombinant DNA

Boyer's work on (*see* Boyer, Herb)
EcoRI identification, 37–38
events surrounding the cleaving of pSC101 with EcoRI, 39–40
introductory information, 31
mapping of alleles of K12 and B, 36
search for recognition site, 36–37
Recombination hotspots, 117
Reese, Colin, 193
Restriction enzymes, 57
Restriction fragment length polymorphism (RFLP), 120
Rett Syndrome
discovery of MeCP2 association with, 17–18
introductory information, 12
possible reversibility, 20
seen as a neuromaintenance disorder, 19
Riggs, Art, 112
Risch, Harvey, 168
Risch, Neil
on the definition of race, 167
family background, 168
introductory information, 164
on the politicizing of the human genome, 167–168
population genetics interest, 164–165
on recruiting more minority scientists into genetics studies, 169
on studying population genetics at Kaiser, 168–169
study of dystonia in Jews, 165–166
Risch, Sonia, 168
RNAi (RNA interference)
Ambros's work on (*see* Ambros, Victor)
anti-sense base pairing determination, 7
determination of an association with *lin-4*, 8–9
introductory information, 1, 2
lin-4 research beginnings, 5–7
RNA splicing
background to, 77–79
Cech's work on (*see* Cech, Tom)
discovery of RNA as a catalyst, 79
introductory information, 75
psoralen cross-linking, 76–77
TERT and, 78–79
Roberts, Rich, 119
Rockefeller Foundation, 90
Roth, John, 28
Ruderman, Joan, 76

Ruvkun, Gary, 5, 7–8, 9
Ruzibakiev, Ruslan, 233, 237
Ryu, Will, 26

Sambrook, Joe, 39
Sanger, Fred, 39, 191
Sarich, Vince, 57
Schena, Mark, 48, 49
Science, 223
Science education
advice for young scientists, 242
Graves' views on kids and science, 110–111
Jeffrey's views on kids and science, 125
legal cases challenging bans on teaching evolution, 131
teaching evolutionary science through books, 73
teaching science to children, 110–111
why students don't become professional scientists, 23–25
Science writing
Carroll's career in, 68, 70, 71–73
Cech on scientists writing about their work, 80
historical approach (*see* History of science)
a journalist's approach (*see* Wade, Nicholas)
on population genetics (*see* Wells, Spencer)
Scopes Trial, 130
Scott, Matt, 64, 65, 67, 76, 190
Shalon, Dari, 46–47
Sharp, Phil, 39, 76, 77, 119
Shubin, Neil, 70
Sinclair, Andrew, 111
Single nucleotide polymorphisms (SNPs), 29
Sir Run Run Shaw Prize, 33
Skolnick, Mark, 28, 29, 30
Smith, Ham, 13, 14, 38
Smithies, Oliver, 176
Sobel, Dava, 91
Solter, Davor, 201
Southern, Ed, 13, 14, 119, 120
Spradling, Al, 76
SRY gene, 111, 114
Staden, Roger, 192
Steitz, Tom, 79
Stem cell therapy, 227. *See also* Embryonic stem cells
Stoneking, Mark, 54, 61, 226

256   Index

Strass, Elaine
  on awards given by GSA, 176
  background as a concert pianist,
    172–173
  duties as executive director, 175
  embryonic lineaging documentation
    example, 184–187
  fondness for her job, 175, 178
  future plans, 178–179
  on how GSA meeting structure works,
    177–178
  introductory information, 172
  jobs leading to position at ASHG, 173
  meetings run by GSA, 173–174
  structure of GSA vs. ASHG, 174–175
  technology improvements for studying
    embryos, 187
Sulston, John
  archiving of e-mails, 89
  comments on future of drug
    development, 196
  comments on his choice of
    study, 188–190
  directorship at Sanger Institute, 192–193
  interest in the interface between science
    and the public, 195–196
  introductory information, 182
  on knighthoods, 197
  nematode genome work, 190–192
  Nobel Prize win, 176
  personal environment needed when
    conducting experiments, 187–188
  on the politicizing of the human genome,
    193–195
  technique used to study live worms,
    183–184
  work with worms, 6, 182–183
Summers, Lawrence, 212
Superconducting supercollider, 228
Supreme Court
  legal definitions used by, 137
  ruling on Dover case, 131
  states' response to creationism
    decision, 132
Swofford, David, 59

Tabby gene, 146
Tabin, Cliff, 70
Telomerase reverse transcriptase
  (TERT), 78–79
Templeton, Alan, 62
Temple University, 216

Thomas Hunt Morgan Award, 176
Thompson, Nichol, 184
Thomson, Jamie
  background to interest in developmental
    biology, 201
  ethical issues with human embryonic
    stem cells, 205
  focus on regenerative medicine, 208
  on the future of human embryonic stem
    cells research, 209
  introductory information, 200
  IPS cell work, 206–208
  move from primate to human embryos,
    204–205
  on the requirement to use existing cell
    lines, 205–206
  success with a new medium for embryos,
    204
  work on deriving primary embryonic
    stem cells, 202–203
  work on deriving primate embryonic
    stem cells, 202, 203–204
Tilghman, Shirley
  Botstein and, 24
  enjoyment of NIH work, 219
  feelings about Princeton, 212–213
  introductory information, 212
  on making career decisions based on
    family obligations, 216–217
  plans for expansion at Princeton,
    213–214
  realization that *H19* is imprinted, 218
  RNA paper, 78
  on selection as University President,
    214–215
  speech on future prospects for
    women in science, 217, 219–220
  teaching duties, 215–216
Tn10, 29
Torsion dystonia, 165–166
Tortoiseshell cat, 148
Trade Related Aspects of Intellectual
  Property Rights (TRIPS), 195
Trask, Barb, 97
Tykocinski, Mark, 15

Ultrabithorax (*Ubx*), 66
unc-86, 5
Underhill, Peter, 238

Varmus, Harold, 205
Venter, Craig, 98, 100, 161, 194

Waddington, C.H., 144, 145
Wade, Nicholas
  background leading to work in science journalism, 223
  book on genetics, 225
  choosing what stories to develop, 226
  ethics of choosing stories, 227–228
  introductory information, 222
  memorable topics, 228
  process used to develop a story, 224–226
  readership, 223
  science stories missed, 227
  on the urgency to write stories, 228
  view of intelligent design, 228
  work as an editorial writer and editor, 223–224
Waldholz, Michael, 28
Wallace, Doug, 60, 73
Wallace, Margaret, 145
Wambaugh, Joseph, 124
Warren, Steve, 96
Waterston, Bob, 187, 192
Watson, Jim, 28, 88, 91, 92
Watt, James, 228
Webster, Matt, 238
Weinberg, Roger, 34
Weinstock, Claire, 45
Weissmann, Charlie, 118
Wellcome Trust Sanger Institute (WTSI), 84, 182, 193
Wells, Spencer
  advice for young scientists, 242
  association with a film production company, 238
  collection of population blood samples, 234–236
  early interest in writing and history, 241
  focus on Central Asia, 233–234
  goal of the Genographic Project, 239
  introductory information, 232
  logistics of second Central Asia trip, 236–237
  public participation response and results, 235, 240–241
  results from second trip, 238
  sampling of extant populations at regional centers, 239
  source of funding for second trip, 237
  transition from academia to field work, 232–233
  use of the Internet to chronicle second trip, 237
West, Mike, 205
White, John, 184, 187
White, Ray, 30, 121
White, Tim, 58, 62
Whitehead Institute, 216
Wieschaus, Eric, 65
Willard, Hunt, 98
Wilmut, Ian, 206
Wilson, Allan, 54, 56, 57, 59, 61, 62, 154, 158
Wisconsin Primate Center, 202
Wistar Institute, 201
Wolpoff, Milford, 60, 62
World Intellectual Property Organization (WIPO), 195
World Trade Organization (WTO), 195
WTSI (Wellcome Trust Sanger Institute), 84, 182, 183
Wyman, Arlene, 30, 121
Wyszewianska, Ania, 27

X chromosome inactivation
  Graves's work on (*see* Graves, Jenny)
  introductory information, 106, 141
  Lyon's work on (*see* Lyon, Mary)
  mottled mice experiments, 146–147
  origin of hypothesis about LINE elements, 148–150
  Tabby mice experiments, 148

Yamanaka, 207
Yoshimori, Bob, 37
Yu, Junying, 206, 207
Yuan, Bob, 36

Zaug, Arthur, 78
Zerjal, Tatiana, 238
ZFY gene, 111, 114
Zhang, Yi, 16
Zoghbi, Huda, 16, 17

About the Author

Jane Gitschier is Professor of Medicine and Pediatrics and a member of the Institute for Human Genetics at the University of California, San Francisco. She received her PhD from MIT in biology and worked as a postdoctoral fellow at Genentech, where she contributed to the cloning of the human gene for coagulation factor VIII to treat hemophiliacs. She has spent her research career at UCSF studying the molecular basis of inherited diseases in man and mouse. She was an investigator with the Howard Hughes Medical Institute for more than 20 years and a Guggenheim Fellow. She is currently investigating the genetic basis for absolute pitch perception and plans to write more on genetics.